The Complete Book of Graphing

by
Douglas C. McBroom

J. WESTON

WALCH
PUBLISHER
Portland, Maine

User's Guide
to
Walch Reproducible Books

As part of our general effort to provide educational materials that are as practical and economical as possible, we have designated this publication a "reproducible book." The designation means that purchase of the book includes purchase of the right to limited reproduction of all pages on which this symbol appears:

Here is the basic Walch policy: We grant to individual purchasers of this book the right to make sufficient copies of reproducible pages for use by all students of a single teacher. This permission is limited to a single teacher and does not apply to entire schools or school systems, so institutions purchasing the book should pass the permission on to a single teacher. Copying of the book or its parts for resale is prohibited.

Any questions regarding this policy or requests to purchase further reproduction rights should be addressed to:

Permissions Editor
J. Weston Walch, Publisher
321 Valley Street • P.O. Box 658
Portland, Maine 04104-0658

1   2   3   4   5   6   7   8   9   10

ISBN 0-8251-3919-8

CONTENTS

TO THE TEACHER

I have always believed that if students could master the basics of graphing, they could graph any equation presented to them. This book was designed with exactly that purpose in mind. It addresses the different areas of graphing in separate sections, so that the teacher can focus on any topics where students need extra work. Each section includes a variety of practice worksheets. With practice, students strengthen their general skills and build strength in specific areas. The reproducible student pages are designed so that students who have a solid grasp of the material will still be challenged and not bored, while lower-level students will not be overwhelmed.

Some teachers choose to use graphing calculators hand-in-hand with teaching students how to graph, while others believe that students should be able to graph first, then use the calculators. This book is designed to support either method of instruction.

This book is intended to be a resource for the teacher, not to actually teach the students how to graph. Instructional pages are included, but they are there to reinforce the teacher's instruction, not replace it. The "Remember" callout balloons on many pages are included to help students remember parts of graphing that they frequently forget.

These materials allso incorporate NCTM's *Standards 2000*. Of course, this book reflects a great deal of the standards listed in Algebra. However, it also covers other areas of the standards, including Numbers and Operations, Geometry, Data Analysis & Probability, Problem Solving, Connections, and Representation.

Name: _____

Date: _____ Period: _____

NUMBER LINES

The number line is an essential concept of algebra. The number line is the basis for constructing one-, two-, or even three-dimensional graphs. Number lines are used in every part of society. Time lines, house numbers, floors of a building, even electrical circuit flowcharts are all forms of number lines. Any system that has a unit of movement or has directed movement is using some form of a number line.

The number line must have four parts: (1) a **line,** (2) an **origin** (this is labeled 0 and is present even if it is not shown in every graph), (3) the **positive side** (on the right of the origin) and the **negative side** (on the left of the origin), and (4) a **constant unit of measurement** between points.

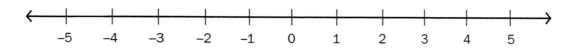

DIRECTIONS: Draw number lines using the given criteria:

1. From −3 to 8

_____

2. From 4 to 15

_____

3. From −10 to 0

_____

4. From −2 to 4 (use 0.5 as the distance between points)

_____

The Complete Book of Graphing

Name: _____

Date: _____ Period: _____

GRAPHING INEQUALITIES

There are six symbols for equality and inequality in algebra. They are:

$=$ *means* equal to	$\neq$ *means* not equal to
$\leq$ *means* less than or equal to	$<$ *means* less than
$\geq$ *means* greater than or equal to	$>$ *means* greater than

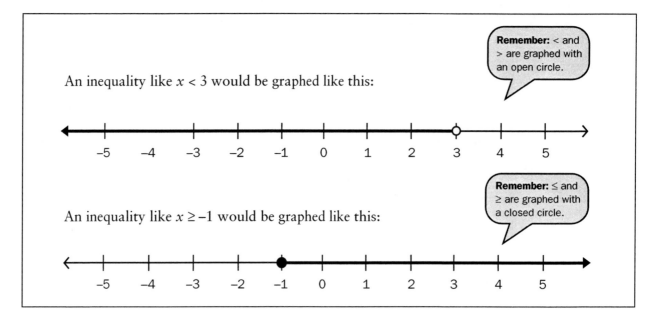

An inequality like $x < 3$ would be graphed like this:

Remember: $<$ and $>$ are graphed with an open circle.

An inequality like $x \geq -1$ would be graphed like this:

Remember: $\leq$ and $\geq$ are graphed with a closed circle.

DIRECTIONS: Graph the following inequalities:

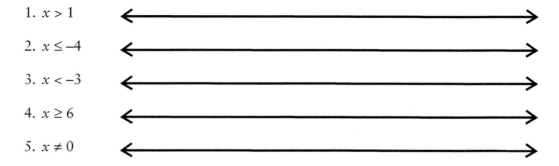

1. $x > 1$

2. $x \leq -4$

3. $x < -3$

4. $x \geq 6$

5. $x \neq 0$

The Complete Book of Graphing

GRAPHING ABSOLUTE VALUE EQUATIONS AND INEQUALITIES

Absolute value equations have two possible solutions. There must be two different setups for each equation or inequality. An example is $|x| = 4$. The solution is that x can be equal to either 4 or −4. If either solution is put into the equation, it will work. The graph of the solution looks like this:

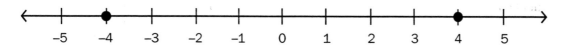

Absolute value inequalities also have two possible solutions. For example, $|x| \leq 2$ will give the solutions $x \leq 2$ and $x \geq -2$. This is called a **conjunction.** The graph looks like this:

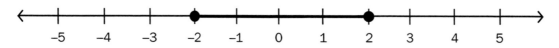

This solution graph shows that any point between −2 and 2 will give a correct solution. An example is −1: $|-1| \leq 2$. Every solution set should always be checked. If the equation is changed to $|x| \geq 2$, this is called a **disjunction,** and the graph looks like this:

Remember: The absolute value of any number **must** be positive.

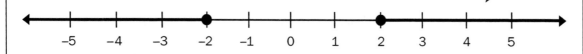

DIRECTIONS: Graph the following:

1. $|x| = 3$ ⟵————————————————————⟶

2. $|x| \leq 4$ ⟵————————————————————⟶

3. $|x| \geq 1$ ⟵————————————————————⟶

4. $|x| > 2$ ⟵————————————————————⟶

5. $|x| < -3$ ⟵————————————————————⟶

Name: _____

Date: _____ Period: _____

SOLVING AND GRAPHING INEQUALITIES

When solving an inequality, the procedure is the same as for solving an equality.

Here is an example of solving an inequality:

Step 1: $3x + 4 < 10$
$\quad\quad\quad\;\; \underline{-4 \quad -4}$

Step 2: $3x + 4 < 10$ $3x < 6$
$\quad\quad\quad\;\; \underline{-4 \quad -4}$

Remember: If you must multiply or divide both sides by a negative number to solve, the direction of the inequality **must** be reversed.

Step 3: $\dfrac{3x}{3} < \dfrac{6}{3}$

SOLUTION: $\quad x < 2$

This is graphed as:

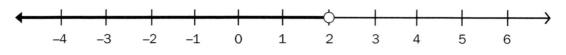

DIRECTIONS: Solve and graph the following inequalities:

1. $2x + 3 > 7$ Solution: _____

2. $5x - 3 \geq 17$ Solution: _____

3. $3x + 4 \leq -5$ Solution: _____

4. $-4x + 2 > 14$ Solution: _____

5. $-3x - 5 < -20$ Solution: _____

The Complete Book of Graphing

Name: _____

Date: _____ Period: _____

SOLVING AND GRAPHING INEQUALITIES

DIRECTIONS: Solve and graph the following inequalities:

1. $2x + 4 > 6$ Solution: _____

2. $3x - 5 \leq 7$ Solution: _____

3. $4x + 6 < -2$ Solution: _____

4. $3x - 6 \geq 0$ Solution: _____

5. $-4x + 2 > 10$ Solution: _____

6. $3x - 6 \leq -6$ Solution: _____

7. $-x + 5 < 1$ Solution: _____

8. $\frac{x}{3} - 2 \geq -3$ Solution: _____

Name: _____

Date: _____ Period: _____

SOLVING AND GRAPHING COMBINED INEQUALITIES

Combined inequalities are two inequalities placed on one graph. The graph can be either a **conjunction** or a **disjunction**. Each inequality is solved in the usual way. Then both solutions are graphed.

DISJUNCTION: $2x + 3 \leq 5$ or $x - 2 \geq 2$ Each inequality is solved normally.

SOLUTIONS: $x \leq 1$ or $x \geq 4$ Both solutions are then graphed.

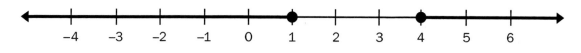

CONJUNCTION: $2x + 3 \geq 5$ and $x - 2 \leq 2$ Each inequality is solved normally.

SOLUTIONS: $x \geq 1$ and $x \leq 4$ Both solutions are graphed, but only the portions that are contained in both solutions. (This is where they overlap.)

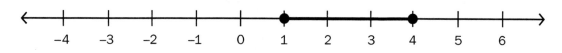

Conjunctions can also be given in this form: $4 < 2x + 6 < 10$

Solving these conjunctions is done as before except that the steps for solving are done twice.

EXAMPLE: $4 < 2x + 6 < 10$ $\begin{array}{c} 4 < 2x + 6 < 10 \\ \underline{-6 \qquad -6 \ -6} \end{array}$ ⟹ $-2 < 2x < 4$

$\dfrac{-2}{2} < \dfrac{2x}{2} < \dfrac{4}{2}$ ⟹ SOLUTION: $-1 < x < 2$

> **Remember:** Always *check* a point in the shaded areas to make sure the graph was done properly.

This solution is graphed as follows:

The Complete Book of Graphing

Name: _____

Date: _____ Period: _____

SOLVING AND GRAPHING COMBINED INEQUALITIES

Solve and graph the following combined inequalities:

> **Remember:** Change signs when multiplying or dividing by a negative.

1. $2x + 4 \leq 0$ or $3x - 2 \geq 7$ Solution(s):_____

2. $3x - 2 < -8$ or $2x - 5 > 3$ Solution(s):_____

3. $5x - 3 \geq 7$ and $2x + 1 \leq 11$ Solution(s):_____

4. $-x \leq 5$ and $-2x + 3 \geq -5$ Solution(s):_____

5. $-4 < 2x < 12$ Solution(s):_____

6. $6 < 3x + 6 < 24$ Solution(s):_____

7. $-8 \leq 2x + 2 \leq 8$ Solution(s):_____

8. $-4 \leq -3x - 1 < 5$ Solution(s):_____

Name: _____

Date: _____ Period: _____

SOLVING AND GRAPHING ABSOLUTE VALUE INEQUALITIES

Absolute value inequalities have two possible solutions, and each inequality must be examined before graphing.

> **Remember:** If IxI < a number, then the graph of the inequality is a **conjunction**. If IxI > a number, then the graph of the inequality is a **disjunction. Don't forget to check your solutions.**

EXAMPLE: $|2x + 3| \leq 5$

The two setups are as follows:

$2x + 3 \leq 5$ and $2x + 3 \geq -5$

Each setup is solved and graphed as before.

Step 1 $2x + 3 \leq 5$ and $2x + 3 \geq -5$
              $-3\ -3$ $-3\ -3$

Step 2 $\dfrac{2x}{2} \leq \dfrac{2}{2}$ and $\dfrac{2x}{2} \geq \dfrac{-8}{2}$

SOLUTIONS: $x \leq 1$ and $x \geq -4$ Another way to write this is: $-4 \leq x \leq 1$

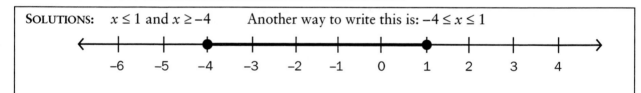

Solve and graph the following absolute value inequalities:

1. $|3x + 6| < 9$ Solutions: _____

2. $|2x - 4| \geq 4$ Solutions: _____

3. $|5x - 10| > 25$ Solutions: _____

Name: _____

Date: _____ Period: _____

SOLVING AND GRAPHING ABSOLUTE VALUE INEQUALITIES

DIRECTIONS: Solve and graph the following:

1. $|2x + 1| > 7$ Solutions: _____

$\longleftarrow\hspace{11cm}\longrightarrow$

2. $|3x - 9| \leq 6$ Solutions: _____

$\longleftarrow\hspace{11cm}\longrightarrow$

3. $|6x + 3| > 9$ Solutions: _____

$\longleftarrow\hspace{11cm}\longrightarrow$

4. $|4x + 2| \leq -4$ Solutions: _____

$\longleftarrow\hspace{11cm}\longrightarrow$

5. $|x - 4| > 0$ Solutions: _____

$\longleftarrow\hspace{11cm}\longrightarrow$

6. $|6x + 3| \leq 12$ Solutions: _____

$\longleftarrow\hspace{11cm}\longrightarrow$

7. $|2x - 5| < 9$ Solutions: _____

$\longleftarrow\hspace{11cm}\longrightarrow$

8. $|5x + 5| > 10$ Solutions: _____

$\longleftarrow\hspace{11cm}\longrightarrow$

Name: _____

Date: _____ Period: _____

MIXED REVIEW OF THE NUMBER LINE

DIRECTIONS: Solve and graph the following:

1. $4x + 6 > 18$ Solution: _____

$$\longleftrightarrow$$

2. $3x - 7 \le -16$ Solution: _____

$$\longleftrightarrow$$

3. $-2x + 5 < -1$ Solution: _____

$$\longleftrightarrow$$

4. $2x - 5 < -3$ or $3x + 1 > -14$ Solutions: _____

$$\longleftrightarrow$$

5. $-4 \le 2x + 8 \le 18$ Solutions: _____

$$\longleftrightarrow$$

6. $-18 < 3x - 3 \le 3$ Solutions: _____

$$\longleftrightarrow$$

7. $|5x - 5| > 25$ Solutions: _____

$$\longleftrightarrow$$

8. $|6x + 3| \le 27$ Solutions: _____

$$\longleftrightarrow$$

Name: _____

Date: _____ Period: _____

THE CARTESIAN PLANE

The **Cartesian plane** (or **coordinate plane**) is a system of four regions or **quadrants** separated by the *x*- and *y*-axes. This is used in everything from maps to architecture to television screens. The *x*-axis is **horizontal,** while the *y*-axis is **vertical.** Below is an example of a typical coordinate plane. The plane has four quadrants, as shown. Notice that the quadrants are labeled counterclockwise.

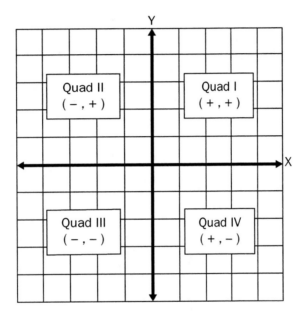

Name: _____

Date: _____ Period: _____

GRAPHING ORDERED PAIRS

Coordinates are written in **ordered pairs** (x,y).

EXAMPLE: (3,2)

To graph ordered pairs, start at the origin, or center. Move the number of units shown by the x-coordinate (also called the abscissa). Go **right** if the number is **positive, left** if it is **negative.** Then go vertically the number of the y-coordinate (the ordinate), **up** if it is **positive, down** if it is **negative.** Using the example of (3,2), first move from the origin right three units, then up two units, as shown. The final position is point (3,2).

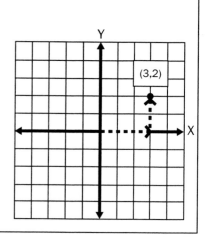

DIRECTIONS: Graph the following points on the graph, connecting them in order.

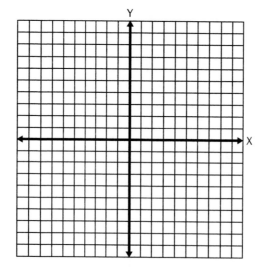

1. (−5,2)
2. (5,2)
3. (−4,−5)
4. (0,6)
5. (4,−5)
6. (−5,2)

The Complete Book of Graphing

Name: _____

Date: _____ Period: _____

GRAPHING ORDERED PAIRS

DIRECTIONS: Graph the following ordered pairs and connect them in order. When the words "Line ends," appear, begin a new line. Go down each column in order.

First Line	Second Line	Third Line
(0,–7)	(0,–4)	(0,–5)
(0,0)	(–1,–4)	(1,–5)
(–1,0)	(–2,–3)	(2,–4)
(–3,2)	(–3,–1)	(4,–2)
(–3,5)	(–2,–2)	(2,–3)
(–2,7)	(0,–4)	(0,–5)
(–1,5)	Line ends	Line ends
(0,7)		
(1,5)		
(2,7)		
(3,5)		
(3,2)		
(1,0)		
(0,0)		
Line ends		

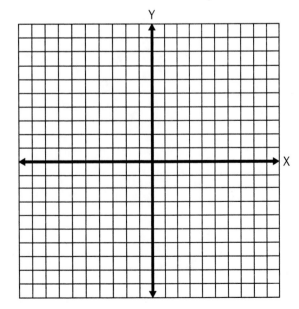

Name: _____

Date: _____ Period: _____

GRAPHING ORDERED PAIRS

DIRECTIONS: Graph the following ordered pairs and connect them in order. When the words "Line ends," appear, begin a new line. Go down each column in order.

First Line	Second Line	Third Line
(–8,0)	(–4,1)	(–4,1)
(–8,1)	(–4,0)	(4,1)
(–7,2)	(–3,1)	Line ends
(–6,3)	(–3,0)	
(–4,4)	(–2,1)	
(–2,5)	(–2,0)	
(0,5)	(–1,1)	
(2,5)	(–1,0)	
(4,4)	(0,1)	
(6,3)	(0,0)	
(7,2)	(1,1)	
(8,1)	(1,0)	
(8,0)	(2,1)	
(7,–1)	(2,0)	
(6,–2)	(3,1)	
(4,–3)	(3,0)	
(2,–4)	(4,1)	
(0,–4)	(4,0)	
(–2,–4)	(–4,0)	
(–4,–3)	Line ends	
(–6,–2)		
(–7,–1)		
(–8,0)		
Line ends		

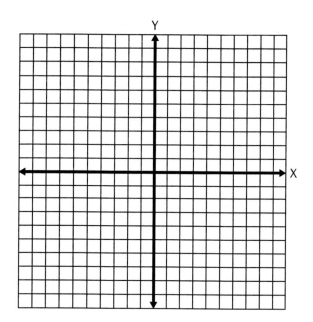

Name: _____

Date: _____ Period: _____

STANDARD FORM AND SLOPE-INTERCEPT FORM OF A LINE

The two most commonly used linear functions are **standard form** and **slope-intercept form.** They appear as follows:

Standard form: $Ax + By = C$

Slope-intercept form: $y = mx + b$

For each equation, the x and y represent the **abscissa** and **ordinate** of a point. If x changes, then y must also change. This forms a **linear function.** In the standard form, the A, B, and C are all integers. In the slope-intercept form, the m and b can be any real number, but the y term must be alone.

> Examples of standard form are:
>
> $2x + 3y = 6$ or $4x - 5y = -12$
>
> Examples of slope-intercept form are:
>
> $y = 2x + 4$ or $y = (-1/2)x - 4$

Remember: It is possible for either the x or y term to be a zero. This means that, in some cases, both coordinates may not be present in the equation.

DIRECTIONS: Identify the following equations as standard form (SF), slope-intercept form (SI), or neither (N):

1. $2x + 5y = 3$ _____

2. $2x + 5 = 10$ _____

3. $y = 6x - 1$ _____

4. $x + 2y = 0$ _____

5. $3y = 2x + 6$ _____

6. $5x - 4 = 2y$ _____

7. $x = 3$ _____

8. $4x + y = -5$ _____

9. $0.5x + 0.2y = 4$ _____

10. $(2/3)x + y = 1$ _____

11. $8x - 11y = 33$ _____

12. $y = 2x$ _____

13. $y = (3/4)x - 2$ _____

14. $3x - 3y = 3$ _____

15. $x = 3y - 9$ _____

16. $2x - 7y = -11$ _____

17. $-4y = x + 1$ _____

18. $y = 3x - (-3)$ _____

19. $y = x$ _____

20. $x + y = 0$ _____

The Complete Book of Graphing

Name: _____

Date: _____ Period: _____

FINDING POINTS ON A LINE

In order to graph a linear equation, **coordinates** must be found. One method of finding points is to use a coordinate table. To find points to graph, simply put values in for either the *x* or *y* value, then solve for the remaining term. This should be repeated at least three times in case a mistake is made. With linear equations, each line must be straight. If the three points do not form a straight line, a mistake has been made and must be corrected.

Here is an example of finding points on a line.

Equation: $2x + y = 3$

The coordinate table will look like this one.
The equation is graphed on the right.

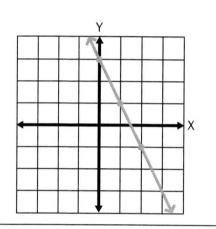

x	y	Coordinate
0	3	(0,3)
1	1	(1,1)
2	−1	(2,−1)

DIRECTIONS: Find three coordinates for each of the following equations:

1. $x + 3y = 6$

x	y	Coordinate

2. $x − y = 1$

x	y	Coordinate

3. $y = (1/2)x − 2$

x	y	Coordinate

4. $y = 3x − 3$

x	y	Coordinate

Name: _____

Date: _____ Period: _____

FINDING POINTS ON A LINE

Remember: Use simple numbers to make the coordinates, and use zero for the *x* and *y* if possible. These are usually easy ordered pairs to find.

DIRECTIONS: Find three coordinates using these equations:

1. $2x - 3y = 6$

x	y	Coordinate

2. $x + y = 3$

x	y	Coordinate

3. $2x + y = -4$

x	y	Coordinate

4. $3x - 6y = -12$

x	y	Coordinate

5. $y = 2x + 4$

x	y	Coordinate

6. $y = (1/3)x - 2$

x	y	Coordinate

7. $y = 5x$

x	y	Coordinate

8. $2x - y = -8$

x	y	Coordinate

Name: _____

Date: _____ Period: _____

FINDING POINTS ON A LINE

> **Remember:** Use simple numbers to make the coordinates, and use zero for the x and y if possible. These are usually easy ordered pairs to find.

DIRECTIONS: Find three points using the following equations:

1. $x - 3y = 0$

x	y	Coordinate

2. $x + y = 2$

x	y	Coordinate

3. $y = 2x - 4$

x	y	Coordinate

4. $y = 5x + 10$

x	y	Coordinate

5. $3x + 3y = -9$

x	y	Coordinate

6. $y = (1/2)x + 3$

x	y	Coordinate

7. $y = x - 1$

x	y	Coordinate

8. $3x + 2y = 1$

x	y	Coordinate

Name: _____

Date: _____ Period: _____

GRAPHING LINES FROM POINTS

DIRECTIONS: Graph the lines of the following equations by finding three points, then making a graph from the points:

Remember: Use small numbers to find coordinates and use multiples of the denominator with fractions.

EXAMPLE: $2x + y = 2$

x	y	Coordinate
0	2	(0,2)
1	0	(1,0)
2	–2	(2,–2)

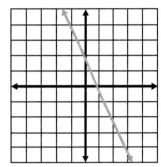

1. $y = x + 3$

x	y	Coordinate

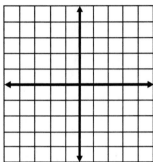

2. $y = (1/2)x - 2$

x	y	Coordinate

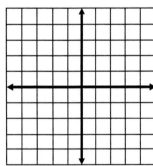

3. $y = 2x$

x	y	Coordinate

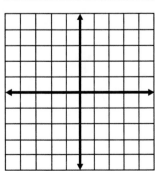

Name: _____

Date: _____ Period: _____

GRAPHING LINES FROM POINTS

DIRECTIONS: Graph the lines of the following equations by finding three points, then making a graph from the points:

1. $x - 2y = -4$

x	y	Coordinate

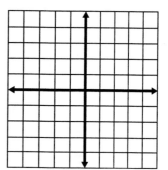

2. $y = x - 1$

x	y	Coordinate

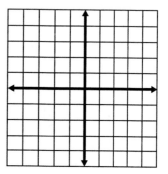

3. $y = (1/3)x + 2$

x	y	Coordinate

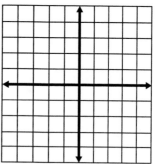

4. $2x + y = 0$

x	y	Coordinate

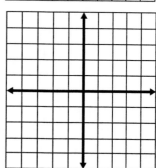

Name: _____

Date: _____ Period: _____

CONVERTING BETWEEN STANDARD FORM AND SLOPE-INTERCEPT FORM

Converting between standard form and slope-intercept form is necessary to effectively graph many equations. The forms are:

standard form $\qquad Ax + By = C$

slope-intercept form $\qquad y = mx + b$

Converting from standard to slope-intercept form requires two steps:

Step 1 Move the x term to the right side of the equation, and change the sign by adding the opposite to both sides.

EXAMPLE: $\qquad 4x - 2y = 6$
$4x + (-4x) - 2y = -4x + 6$
$-2y = -4x + 6$

Step 2 Divide everything by the y coefficient.

$$\frac{-2y}{-2} = \frac{-4x}{-2} + \frac{6}{-2}$$

Remember: Always simplify fractions.

The final answer becomes: $\qquad y = 2x - 3$

Convert the following equations from standard to slope-intercept form:

1. $2x - 3y = 9$ _____

2. $x + 2y = -4$ _____

3. $3x - y = 7$ _____

4. $5x + 3y = 9$ _____

5. $2x - 2y = 2$ _____

6. $x - 4y = 6$ _____

7. $3x + 2y = 0$ _____

8. $4x - 3y = 6$ _____

9. $x - y = 4$ _____

10. $4x + y = 1$ _____

Name: _____

Date: _____ Period: _____

CONVERTING BETWEEN STANDARD FORM AND SLOPE-INTERCEPT FORM

Converting from slope-intercept to standard form requires two steps:

Step 1 Move the x term to the left side of the equation, and change the sign by adding the opposite to both sides.

EXAMPLE:	
	$y = (2/3)x - 1$
	$-(2/3)x + y = (2/3)x - (2/3)x - 1$
	$\dfrac{-2x}{3} + y = -1$

Step 2 Multiply everything by the least common denominator.

$$3\frac{(-2x)}{3} + 3(y) = 3(-1)$$

$$-2x + 3y = -3$$

If the x term is negative, multiply everything by a (-1).

$$(-1)(-2x) + (-1)(3y) = (-1)(-3)$$

The final answer becomes: $2x - 3y = 3$

> **Remember:** In standard form each term must be an integer and the x-term is positive.

Convert the following from slope-intercept to standard form:

1. $y = 2x - 3$ _____

2. $y = 3x + 1$ _____

3. $y = -4x + 5$ _____

4. $y = -x - 2$ _____

5. $y = (2/3)x - 4$ _____

6. $y = (-1/4)x + 1$ _____

7. $y = -2x$ _____

8. $y = (4/3)x - 3$ _____

9. $y = (-5/3)x + 10$ _____

10. $y = -3x - 3$ _____

Name: _____

Date: _____ Period: _____

GRAPHING LINEAR EQUATIONS

The simplest method of graphing a linear equation is from the slope-intercept form. From this form, $y = mx + b$, the y-intercept and the slope of the line are given. There are only two steps:

Step 1 The b is the y-intercept, which is the starting point. Locate this point on the y-axis.

Step 2 Find points on the line by using the slope m from the equation. The slope must be in fractional form if it is not already.

EXAMPLE: $y = 2x - 1$ $m = 2$ (this becomes 2/1 in fractional form) $b = -1$

The starting point of the graph is (0,–1) from the y-intercept. The line's slope is 2/1 (both are positive so the line will go up two units as it goes right one unit). The example looks like this:

Remember: Slope moves vertically (the numerator), then horizontally (the denominator). **Up** and **right** are **positive**. **Down** and **left** are **negative**.

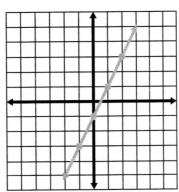

---

1. Graph: $y = 3x - 2$

   $m =$ _____
   $b =$ _____

2. Graph: $y = (1/2)x + 1$

   $m =$ _____
   $b =$ _____

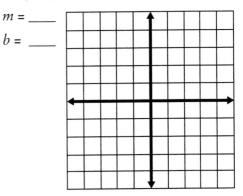

The Complete Book of Graphing

Name: _____

Date: _____ Period: _____

GRAPHING LINEAR EQUATIONS

DIRECTIONS: Find the slope and *y*-intercept, then graph each equation:

1. $y = x + 2$

   $m =$ _____ $b =$ _____

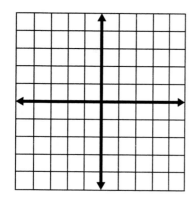

2. $y = 2x - 3$

   $m =$ _____ $b =$ _____

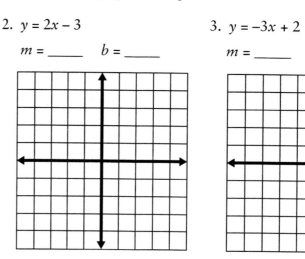

3. $y = -3x + 2$

   $m =$ _____ $b =$ _____

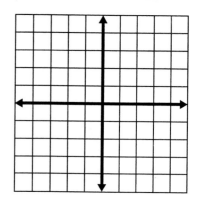

4. $y = (1/2)x - 1$

   $m =$ _____ $b =$ _____

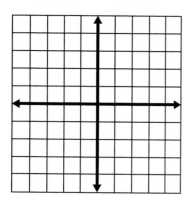

5. $y = (-2/3)x + 2$

   $m =$ _____ $b =$ _____

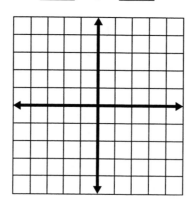

6. $y = 2x + (1/2)$

   $m =$ _____ $b =$ _____

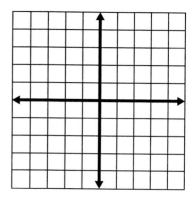

Name: _____

Date: _____ Period: _____

GRAPHING LINEAR EQUATIONS

DIRECTIONS: Convert the following linear equations to slope-intercept (SI) form, then find the slope and the *y*-intercept and graph the equation.

1. $6x - 3y = 9$

SI: _____

$m =$ _____ $b =$ _____

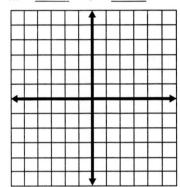

2. $2x + y = 4$

SI: _____

$m =$ _____ $b =$ _____

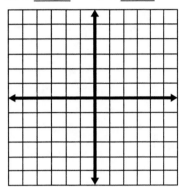

3. $2x - 3y = -6$

SI: _____

$m =$ _____ $b =$ _____

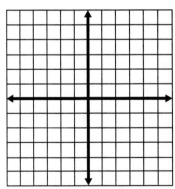

4. $3x + y = 0$

SI: _____

$m =$ _____ $b =$ _____

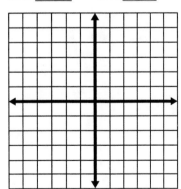

Name: _____

Date: _____ Period: _____

FINDING SLOPE FROM TWO POINTS

The slope (or steepness) of a line is found with a simple formula that uses the change in the vertical position divided by the change in the horizontal position.

The formula is usually seen as:

$$m = \frac{\text{rise}}{\text{run}} = \frac{y_2 - y_1}{x_2 - x_1}$$

Using the points (3,2) and (5,6), the slope would be as follows:
Label the points:

$$x_1 = 3, \; y_1 = 2, \; x_2 = 5, \; y_2 = 6$$

The formula would be:

$$m = \frac{y_2 - y_1}{x_2 - x_1} = \frac{6 - 2}{5 - 3} = \frac{4}{2} = 2$$

So, the slope is 2, which can also be written as $m = 2$.

Remember: The x- and y-terms can be reversed. Either abscissa can be x_1 or x_2. However, the ordinate must stay with its respective abscissa. You cannot mix the two coordinates.

The slope must be in the form of an **integer**, a **proper fraction**, or an **improper fraction**. Do not use mixed numbers or reducible fractions. Watch for sign changes.

DIRECTIONS: Find the slope of the line that would contain the following points:

1. (1,4), (2,2) $m =$ _____

2. (5,1), (1,3) $m =$ _____

3. (3,–2), (4,1) $m =$ _____

4. (–2,–4), (–1,3) $m =$ _____

5. (3,0), (–2,3) $m =$ _____

6. (–4,2), (4,–2) $m =$ _____

7. (0,0), (3,5) $m =$ _____

8. (5,–1), (–1,–3) $m =$ _____

9. (–6,7), (4,2) $m =$ _____

10. (–4,3), (2,3) $m =$ _____

The Complete Book of Graphing

Name: _____

Date: _____ Period: _____

FINDING SLOPE FROM TWO POINTS

DIRECTIONS: Find the slope of the line that would contain the following points:

1. $(3,2), (4,1)$ $m =$ _____

2. $(1,5), (0,3)$ $m =$ _____

3. $(4,-3), (-1,6)$ $m =$ _____

4. $(7,-4), (-1,-3)$ $m =$ _____

5. $(5,2), (-3,-5)$ $m =$ _____

6. $(0,0), (5,2)$ $m =$ _____

7. $(-3,-4), (-1,-6)$ $m =$ _____

8. $(4,3), (-3,-2)$ $m =$ _____

9. $(-1,0), (1,1)$ $m =$ _____

10. $(0,2), (4,10)$ $m =$ _____

11. $(0,1), (5,-3)$ $m =$ _____

12. $(2,3), (-1,-2)$ $m =$ _____

13. $(12,8), (-4,-10)$ $m =$ _____

14. $(10,-5), (6,-2)$ $m =$ _____

15. $(5,-3), (-4,1)$ $m =$ _____

16. $(4,-7), (0,0)$ $m =$ _____

17. $(-4,1), (4,-1)$ $m =$ _____

18. $(0,6), (8,3)$ $m =$ _____

19. $(-1,-3), (1,3)$ $m =$ _____

20. $(5,1), (-3,2)$ $m =$ _____

Name: _____

Date: _____ Period: _____

FINDING SLOPE FROM A POINT AND THE Y-INTERCEPT

To find the slope of a line with one point and the y-intercept, use the y-intercept as the second point. Then, use the same procedure as though two points were provided.

EXAMPLE:	Point $(2,4)$ and intercept $(b) = 3$, which becomes $(0,3)$
	$$m = \frac{y_2 - y_1}{x_2 - x_1} = \frac{3-4}{0-2} = \frac{-1}{-2} = \frac{1}{2}$$
	So the slope would be 1/2.

DIRECTIONS: Find the slope of the line that would contain the following points:

1. $(3,1)$ $b = 2$ $m =$ _____

2. $(1,4)$ $b = 1$ $m =$ _____

3. $(-1,3)$ $b = -2$ $m =$ _____

4. $(-2,-2)$ $b = -5$ $m =$ _____

5. $(-1,5)$ $b = 3$ $m =$ _____

6. $(-3,1)$ $b = 2$ $m =$ _____

7. $(2,0)$ $b = 4$ $m =$ _____

8. $(1,1)$ $b = -5$ $m =$ _____

9. $(1,-3)$ $b = -3$ $m =$ _____

10. $(2,-3)$ $b = 0$ $m =$ _____

11. $(3,-2)$ $b = 3$ $m =$ _____

12. $(1,3)$ $b = 2$ $m =$ _____

13. $(5,2)$ $b = -1$ $m =$ _____

14. $(-5,-3)$ $b = -3$ $m =$ _____

15. $(4,-2)$ $b = 8$ $m =$ _____

Name: _____

Date: _____ Period: _____

NO SLOPE AND ZERO SLOPE

Some lines have **no slope** or have an **undefined slope.** This means that the line is **vertical.** There is no slope because if the line is vertical, then each x term will be the same. When these terms are put into the slope formula, the **denominator** is **zero,** which is undefined.

Also, lines can have a **zero slope.** These lines are **horizontal.** There is a slope of zero because the y terms are all the same. When these are put into the slope formula, the **numerator** is **zero.** Thus, the line has a zero slope. The following are examples of both:

EXAMPLE OF NO SLOPE: $(3,2)(3,-1)$

$$m = \frac{y_2 - y_1}{x_2 - x_1} = \frac{-1 - 2}{3 - 3} = \frac{-3}{0} = \text{no slope}$$

EXAMPLE OF ZERO SLOPE: $(2,3)(-1,3)$

$$m = \frac{y_2 - y_1}{x_2 - x_1} = \frac{3 - 3}{-1 - 2} = \frac{0}{-3} = 0 \text{ slope}$$

Remember: If the slope is **zero,** then it is written as **"0"** because it still has a slope. If the slope is **undefined,** it is written as **"no slope."**

DIRECTIONS: Find the slope of the line containing the following points:

1. $(2,1), (-1,1)$ $m =$ _____

2. $(3,-2), (3,2)$ $m =$ _____

3. $(-1,1), (1,-1)$ $m =$ _____

4. $(5,2), (3,2)$ $m =$ _____

5. $(-1,4), (-1,2)$ $m =$ _____

6. $(3,2), (-3,-2)$ $m =$ _____

7. $(2,-1), (-2,1)$ $m =$ _____

8. $(3,5), (3,-5)$ $m =$ _____

9. $(1,0), (3,0)$ $m =$ _____

10. $(-2,-2), (3,-2)$ $m =$ _____

Name: _____

Date: _____ Period: _____

FINDING SLOPE FROM A GRAPH

To find slope from a graph, two points must be found. The easiest points are where a line and an axis intersect. After two points have been identified, use the slope formula with these points.

Slope formula: $m = \dfrac{\text{rise}}{\text{run}}$

EXAMPLE: Two points on this gray line are $(-1,-1)$ and $(0,1)$. Other points could be chosen, but choose points that are easy to work with. The slope is written as

$$m = \frac{\text{rise}}{\text{run}} = \frac{y_2 - y_1}{x_2 - x_1} = \frac{1 - (-1)}{0 - (-1)} = \frac{2}{1} = 2$$

The **rise** is 2 units up for every 1 unit to the right, which is the **run**.

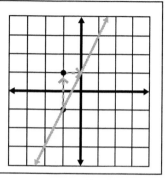

> **Remember: Up** and **right** are both **positive** for rise and run. **Down** and **left** are both **negative** for rise and run.

DIRECTIONS: Find the slope of the following gray lines:

1. $m =$ _____

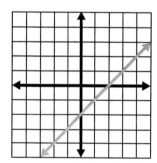

2. $m =$ _____

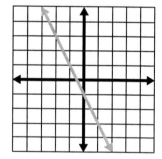

3. $m =$ _____

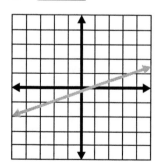

4. $m =$ _____

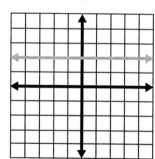

The Complete Book of Graphing

Name: _____

Date: _____ Period: _____

FINDING SLOPE FROM A GRAPH

DIRECTIONS: Find the slope of each gray line below:

1.   *m* = _____

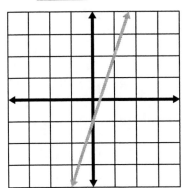

2.   *m* = _____

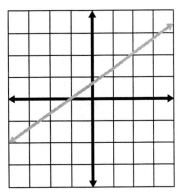

3.   *m* = _____

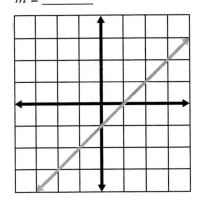

4.   *m* = _____

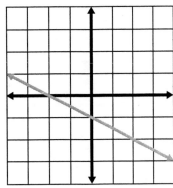

5.   *m* = _____

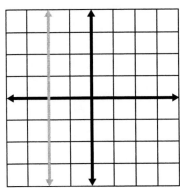

6.   *m* = _____

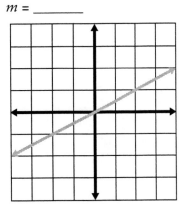

Name: _____

Date: _____ Period: _____

FINDING SLOPE FROM A GRAPH

DIRECTIONS: Find the slope of each gray line below:

1. *m* = _____

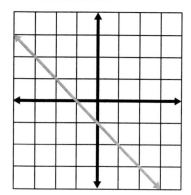

2. *m* = _____

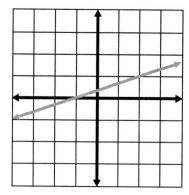

3. *m* = _____

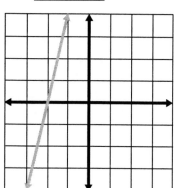

4. *m* = _____

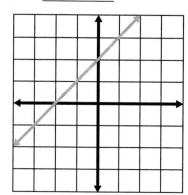

5. *m* = _____

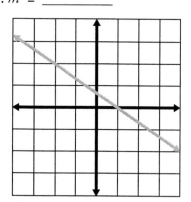

6. *m* = _____

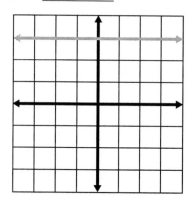

Name: _____

Date: _____ Period: _____

FINDING SLOPE FROM POINTS OR A GRAPH

DIRECTIONS: Find the slope of the line that contains the following points:

1. (3,4), (2,6) $m =$ _____

2. (−3,5), (3,1) $m =$ _____

3. (−5,−5), (5,−5) $m =$ _____

4. (0,−2), (3,−6) $m =$ _____

5. (7,−1), (−3,2) $m =$ _____

6. (−2,−4), (−1,−6) $m =$ _____

DIRECTIONS: Find the slope of each gray line below:

7. $m =$ _____

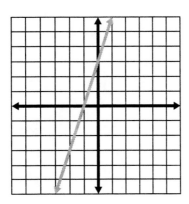

8. $m =$ _____

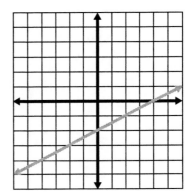

9. $m =$ _____

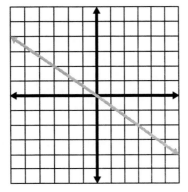

10. $m =$ _____

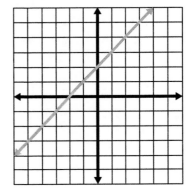

The Complete Book of Graphing

Name: _____

Date: _____ Period: _____

FINDING THE INTERCEPTS FROM A GRAPH

Remember: The intercepts must be ordered pairs such as (2,0) or (0,–1).

DIRECTIONS: Find the x- and y-intercepts from the following graphs:

1. x-intercept = (,   )
   y-intercept = (,   )

2. x-intercept = (,   )
   y-intercept = (,   )

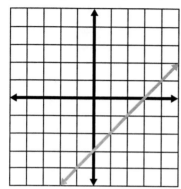
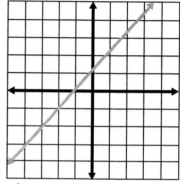

3. x-intercept = (,   )
   y-intercept = (,   )

4. x-intercept = (,   )
   y-intercept = (,   )

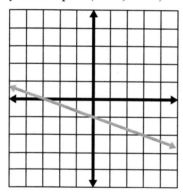
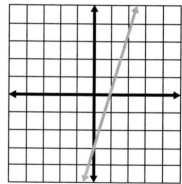

5. x-intercept = (,   )
   y-intercept = (,   )

6. x-intercept = (,   )
   y-intercept = (,   )

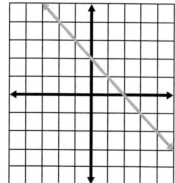
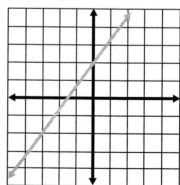

The Complete Book of Graphing

Name: _____

Date: _____ Period: _____

MIXED REVIEW OF LINEAR FUNCTIONS AND SLOPE

DIRECTIONS: Find three points for each of the following equations:

1. $2x - 4y = 8$ 2. $y = 3x - 2$

x	y	Coordinate

x	y	Coordinate

DIRECTIONS: Convert the following equations from standard form to slope-intercept form:

3. $2x + 3y = 6$ _____

4. $3x - y = -2$ _____

DIRECTIONS: Convert the following equations from slope-intercept form to standard form:

5. $y = 2x - 3$ _____

6. $y = (-3/4)x + 1$ _____

DIRECTIONS: Find the slope from the following pairs of points or from the *y*-intercept and a point:

7. $(3,4), (-1,3)$ $m =$ _____

8. $(0,-1), (-2,-4)$ $m =$ _____

9. $(-2,2), b = -1$ $m =$ _____

10. $(3,-4), b = 5$ $m =$ _____

11. $(2,0), (2,-4)$ $m =$ _____

12. $(-4,-4), b = -4$ $m =$ _____

Name: _____

Date: _____ Period: _____

FINDING THE EQUATION OF A LINE FROM THE SLOPE AND THE *Y*-INTERCEPT

To find the equation of a line from the slope and the *y*-intercept simply put in the terms for the variables in the equation.

EXAMPLE: $m = 3$ $b = -2$

Put in 3 and −2 for their respective variables in the equation $y = mx + b$.

The equation would look like this:

$$y = 3x + (-2)$$

Or it could look like this:

$$y = 3x - 2$$

Either form is acceptable since + (−2) is equivalent to −2.

Remember: Once the *m* and *b* are in the equation, there is really nothing else to do. Don't overwork the equation.

DIRECTIONS: Write the equation of each line below in slope-intercept form:

1. $m = 4$ $b = 3$ slope intercept form: _____

2. $m = -3$ $b = -1$ slope intercept form: _____

3. $m = 1$ $b = 12$ slope intercept form: _____

4. $m = (1/3)$ $b = -3$ slope intercept form: _____

5. $m = -2$ $b = 0$ slope intercept form: _____

6. $m = (-2/5)$ $b = 4$ slope intercept form: _____

7. $m = 0$ $b = 3$ slope intercept form: _____

8. $m = (4/3)$ $b = (4/5)$ slope intercept form: _____

9. $m = -6$ $b = 1$ slope intercept form: _____

10. $m = 25$ $b = 75$ slope intercept form: _____

11. $m = (-7/3)$ $b = -6$ slope intercept form: _____

12. $m = 0$ $b = 0$ slope intercept form: _____

Name: _____

Date: _____ Period: _____

FINDING THE EQUATION OF A LINE FROM THE SLOPE AND A POINT

To find the equation of a line from the slope (m) and a point (P), use the slope-intercept form, substitute the given terms, and then solve for the y-intercept (b).

> **EXAMPLE:** $m = 2$ P(4,3)

Remember: Multiply the slope and abscissa, then subtract the number from both sides to find the y-intercept (b).

Step 1: Put the terms into their respective places in the slope-intercept form, $y = mx + b$.

$$3 = 2(4) + b \implies 3 = 8 + b$$

Step 2: Solve for b.

$$3 + (-8) = 8 + (-8) + b \implies -5 = b$$

Step 3: Use the slope and the y-intercept to complete the equation.

$$y = 2x + (-5) \text{ or } y = 2x - 5$$

DIRECTIONS: Find the equation of each line below, then write it in slope-intercept form:

1. $m = 4$ P(1,3)   slope-intercept form: _____

2. $m = -2$ P(2,4)   slope-intercept form: _____

3. $m = 5$ P(-1,-3)   slope-intercept form: _____

4. $m = 3$ P(-2,4)   slope-intercept form: _____

5. $m = -1$ P(-2,0)   slope-intercept form: _____

6. $m = (3/4)$ P(8,5)   slope-intercept form: _____

7. $m = (-1/2)$ P(4,6)   slope-intercept form: _____

8. $m = (5/3)$ P(0,2)   slope-intercept form: _____

9. $m = 5$ P(2,-2)   slope-intercept form: _____

10. $m = -3$ P(1,1)   slope-intercept form: _____

FINDING THE EQUATION OF A LINE USING POINT-SLOPE FORM

The point-slope form of a line is similar to the slope-intercept form. This is what it looks like:

$$y - y_1 = m(x - x_1)$$

The y_1 and x_1 terms are for the x and y coordinates. Simply insert the x and y coordinates and the slope into the equation; then, using solving techniques, put the equation into slope-intercept or standard form.

EXAMPLE: $m = 2$ P(1,4)

$$y - y_1 = m(x - x_1)$$
$$y - 4 = 2(x - 1)$$
$$y - 4 = 2x - 2$$
$$y - 4 + 4 = 2x - 2 + 4$$

Thus the equation is: $y = 2x + 2$

DIRECTIONS: Find the equation of the following lines (in slope-intercept form) using point-slope form:

1. $m = 3$ P(2,3)   slope-intercept form: _____

2. $m = 2$ P(1,4)   slope-intercept form: _____

3. $m = -3$ P(-2,3)   slope-intercept form: _____

4. $m = -2$ P(3,-6)   slope-intercept form: _____

5. $m = 4$ P(0,1)   slope-intercept form: _____

6. $m = -4$ P(-3,-2)   slope-intercept form: _____

7. $m = 0$ P(-7,4)   slope-intercept form: _____

8. $m = (1/2)$ P(4,0)   slope-intercept form: _____

9. $m = -1$ P(5,-5)   slope-intercept form: _____

10. $m = (-3/2)$ P(4,1)   slope-intercept form: _____

11. $m = -4$ P(-2,4)   slope-intercept form: _____

12. $m = (-2/3)$ P(3,-6)   slope-intercept form: _____

Name: _____

Date: _____ Period: _____

FINDING THE EQUATION OF A LINE FROM A POINT AND THE *Y*-INTERCEPT

To find the equation of a line from a point and the *y*-intercept, the slope must be found. Each term is put into its respective place in the slope-intercept form and then the slope (*m*) is found. You can then use the slope and *y*-intercept in the equation.

EXAMPLE: P(3,1)   *y*-intercept = –5

$$y = mx + b \implies 1 = m(3) + (-5)$$

$$1 + 5 = 3m + (-5) + 5 \implies 6 = 3m$$

$$\frac{6}{3} = \frac{3m}{3} \implies m = 2$$

The equation would look like this: $y = 2x + (-5)$ or $y = 2x - 5$

An additional method of finding the slope is to use the slope formula and use the *y*-intercept as a point with the given coordinate.

EXAMPLE: P(3,1)   *y*-intercept = –5   the *y*-intercept is the point (0,–5)

$$m = \frac{y_2 - y_1}{x_2 - x_1}$$

$$m = \frac{-5 - 1}{0 - 3} = \frac{-6}{-3} = 2$$

$$m = 2$$

DIRECTIONS: Find the slope from the given points and *y*-intercepts, then write the SI form of the line:

1. P(1,2)   *y*-intercept = 3   *m* = _____   SI = _____

2. P(–2,–1)   *y*-intercept = 1   *m* = _____   SI = _____

3. P(–3,4)   *y*-intercept = 2   *m* = _____   SI = _____

4. P(1,0)   *y*-intercept = –2   *m* = _____   SI = _____

5. P(5,–4)   *y*-intercept = 1   *m* = _____   SI = _____

6. P(–4,0)   *y*-intercept = 3   *m* = _____   SI = _____

Name: _____

Date: _____ Period: _____

FINDING THE EQUATION OF A LINE FROM TWO POINTS

When the equation of a line must be found from two points, the slope must be found first, then the intercept. Use the slope formula to find the slope, then use the slope and a point in the slope-intercept form to find the y-intercept (b).

EXAMPLE: $P_1(2,1)$ $P_2(4,5)$

Find the slope:

$$m = \frac{y_2 - y_1}{x_2 - x_1} = \frac{5 - 1}{4 - 2} = \frac{4}{2} = 2$$

$$m = 2$$

Find the y-intercept: (either point can be used)

$$y = mx + b \implies 1 = 2(2) + b$$
$$1 = 4 + b$$
$$1 - 4 = 4 - 4 + b$$
$$b = -3$$

Write the equation: $y = mx + b$

$$y = 2x + (-3) \text{ or } y = 2x - 3$$

DIRECTIONS: Use the given points to find the slope, y-intercept, and equation of a line:

1. $P_1(1,3)$ $P_2(0,1)$ $m =$ _____ $b =$ _____   SI: _____

2. $P_1(-2,1)$ $P_2(1,4)$ $m =$ _____ $b =$ _____   SI: _____

3. $P_1(3,-4)$ $P_2(6,2)$ $m =$ _____ $b =$ _____   SI: _____

4. $P_1(0,-2)$ $P_2(4,-4)$ $m =$ _____ $b =$ _____   SI: _____

5. $P_1(-3,-1)$ $P_2(-1,5)$ $m =$ _____ $b =$ _____   SI: _____

6. $P_1(5,-2)$ $P_2(2,1)$ $m =$ _____ $b =$ _____   SI: _____

7. $P_1(4,4)$ $P_2(-4,2)$ $m =$ _____ $b =$ _____   SI: _____

8. $P_1(-1,5)$ $P_2(4,4)$ $m =$ _____ $b =$ _____   SI: _____

9. $P_1(0,0)$ $P_2(2,3)$ $m =$ _____ $b =$ _____   SI: _____

10. $P_1(-2,-3)$ $P_2(2,3)$ $m =$ _____ $b =$ _____   SI: _____

Name: _____

Date: _____ Period: _____

FINDING THE EQUATION OF A LINE FROM TWO POINTS

DIRECTIONS: Write the equation of the line containing the following points in slope-intercept form:

EXAMPLE: $P_1(2,1)$	$P_2(3,0)$	SI:	$y = -x + 3$

1. $P_1(4,1)$ $P_2(2,5)$ SI: _____

2. $P_1(-1,3)$ $P_2(2,6)$ SI: _____

3. $P_1(-2,-2)$ $P_2(0,-4)$ SI: _____

4. $P_1(0,5)$ $P_2(-1,4)$ SI: _____

5. $P_1(0,0)$ $P_2(2,1)$ SI: _____

6. $P_1(-3,-2)$ $P_2(-1,4)$ SI: _____

7. $P_1(2,-1)$ $P_2(0,-2)$ SI: _____

8. $P_1(3,5)$ $P_2(-3,1)$ SI: _____

9. $P_1(2,2)$ $P_2(-1,5)$ SI: _____

DIRECTIONS: Write the equation of the line containing the following points in standard form:

EXAMPLE: $P_1(2,3)$	$P_2(1,5)$	standard form:	$2x + y = 7$

10. $P_1(2,-1)$ $P_2(1,2)$ standard form: _____

11. $P_1(-1,-3)$ $P_2(2,0)$ standard form: _____

12. $P_1(-4,2)$ $P_2(-2,3)$ standard form: _____

13. $P_1(3,-1)$ $P_2(0,0)$ standard form: _____

14. $P_1(-1,-1)$ $P_2(1,0)$ standard form: _____

15. $P_1(3,-2)$ $P_2(6,-3)$ standard form: _____

16. $P_1(-4,2)$ $P_2(-3,4)$ standard form: _____

FINDING THE EQUATION OF A LINE WITH ZERO OR NO SLOPE

To find the equation of a line with zero or no slope, the slope formula must first be used to find the slope. The slope will be either 0 or no slope. The slope will determine the equation of the line. A slope of 0 results in the equation

$$y = \text{(the } y\text{-intercept)},$$

while no slope results in the equation

$$x = \text{(the } x\text{-intercept)}.$$

EXAMPLE 1:	EXAMPLE 2:
$P_1(3,1)$ $P_2(-2,1)$	$P_1(4,-2)$ $P_2(4,5)$
$m = \dfrac{x_2 - y_1}{x_2 - y_1} = \dfrac{1-1}{-2-3} = 0$	$m = \dfrac{x_2 - y_1}{x_2 - y_1} = \dfrac{5-(-2)}{4-4} = \dfrac{7}{0} = \text{undefined}$
The y-intercept is whatever the y is in each coordinate. In this example, the y is 1. The equation is:	The x-intercept is whatever the x is in each coordinate. In this example, the x is 4. The equation is:
$$y = 1$$	$$x = 4$$
This line is horizontal.	This line is vertical.

DIRECTIONS: Find the equation of the following lines with zero or no slope:

1. $P_1(1,3)$ $P_2(-4,3)$ $m = $ _____   equation: _____

2. $P_1(-3,4)$ $P_2(-3,0)$ $m = $ _____   equation: _____

3. $P_1(5,2)$ $P_2(5,-2)$ $m = $ _____   equation: _____

4. $P_1(-1,0)$ $P_2(-1,6)$ $m = $ _____   equation: _____

5. $P_1(3,-3)$ $P_2(-3,-3)$ $m = $ _____   equation: _____

6. $P_1(4,-2)$ $P_2(-2,-2)$ $m = $ _____   equation: _____

Name: _____

Date: _____ Period: _____

MIXED REVIEW OF LINEAR EQUATIONS

DIRECTIONS: Find the equation of each line using the slope and *y*-intercept, then write it in slope-intercept form:

    1. $m = 3$ $b = -5$ slope-intercept form: _____

    2. $m = -2$ $b = 3$ slope-intercept form: _____

    3. $m = -1$ $b = 6$ slope-intercept form: _____

    4. $m = (1/2)$ $b = -1$ slope-intercept form: _____

    5. $m = (-4/3)$ $b = (5/2)$ slope-intercept form: _____

DIRECTIONS: Find the equation of each line using the slope and a point, then write it in slope-intercept form:

    6. $m = 3$ P(2,2)   slope-intercept form: _____

    7. $m = -2$ P(3,–1)   slope-intercept form: _____

    8. $m = 5$ P(–3,–5)   slope-intercept form: _____

    9. $m = (3/4)$ P(8,–1)   slope-intercept form: _____

  10. $m = (-1/2)$ P(6,3)   slope-intercept form: _____

DIRECTIONS: Find the equation of each line using two points, then write it in slope-intercept form:

  11. $P_1(2,1)$ $P_2(0,3)$ slope-intercept form: _____

  12. $P_1(5,2)$ $P_2(-1,-4)$ slope-intercept form: _____

  13. $P_1(-2,3)$ $P_2(1,-3)$ slope-intercept form: _____

  14. $P_1(0,-4)$ $P_2(4,-2)$ slope-intercept form: _____

  15. $P_1(2,-3)$ $P_2(0,0)$ slope-intercept form: _____

  16. $P_1(3,5)$ $P_2(-3,5)$ slope-intercept form: _____

  17. $P_1(-1,0)$ $P_2(-1,4)$ slope-intercept form: _____

  18. $P_1(-3,-4)$ $P_2(3,0)$ slope-intercept form: _____

Name: _____

Date: _____ Period: _____

FINDING THE EQUATION OF A LINE FROM A GRAPH

To write the equation of a line from a graph, two points must first be found. Use these points to find the slope of the line, and from there use either point to find the y-intercept. If possible, one point should be where the line crosses the y-axis, since this point will be the y-intercept.

EXAMPLE: For this line, two points could be (2,3) and (0,–1). First the slope must be found.

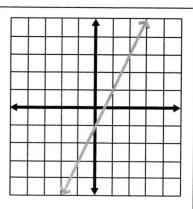

$$m = \frac{y_2 - y_1}{x_2 - x_1} = \frac{-1-3}{0-2} = \frac{-4}{2} = 2$$

$$m = 2$$

To find the y-intercept, choose a point and substitute it into the slope-intercept form to find b. Using point (2,3), the equation would be as follows:

$$y = mx + b \implies 3 = 2(2) + b \implies 3 = 4 + b$$
$$3 + (-4) = 4 + (-4) + b$$
$$b = -1$$

The equation would be:

$$y = 2x - 1$$

Remember: Any two points can be used, so try to choose points that are simple. **Note:** the y-intercept ($b = -1$) was the second point that was chosen. This is a shortcut that can be used at times.

DIRECTIONS: Find the equation of the line from the graph (in slope-intercept form):

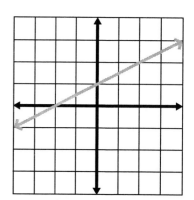

$m =$ _____

$b =$ _____

Equation: _____

Name: _____

Date: _____ Period: _____

FINDING THE EQUATION OF A LINE FROM A GRAPH

DIRECTIONS: Find the equation of the line of each graph (in slope-intercept form):

1. Equation: _____

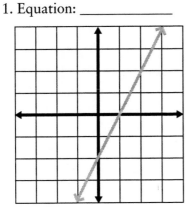

2. Equation: _____

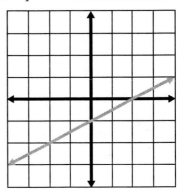

3. Equation: _____

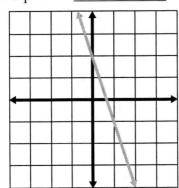

4. Equation: _____

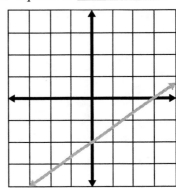

5. Equation: _____

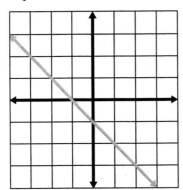

6. Equation: _____

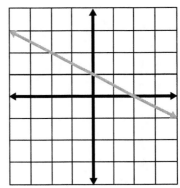

The Complete Book of Graphing

Name: _____

Date: _____ Period: _____

GRAPHING LINEAR EQUATIONS

DIRECTIONS: Graph the following linear equations:

1. $y = -2x + 3$

2. $y = (3/4)x - 2$

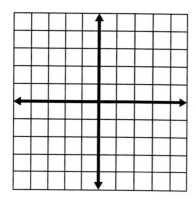

3. $3x + y = -1$

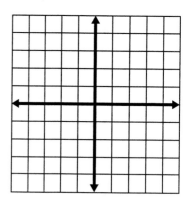

4. $4x - 2y = 8$

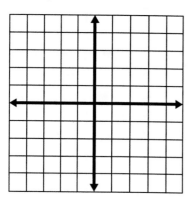

5. $4x + 8y = -8$

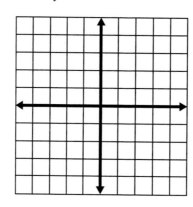

6. $x - y = -3$

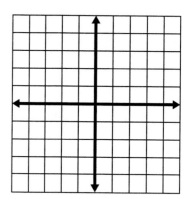

7. $x + 3y = 9$

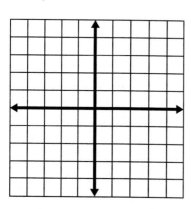

8. $4x + y = 0$

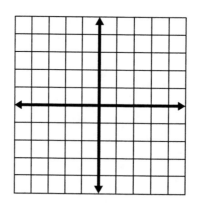

The Complete Book of Graphing

Name: _____

Date: _____ Period: _____

GRAPHING LINEAR EQUATIONS

DIRECTIONS: Graph the following linear equations:

1. $y = 3x - 2$

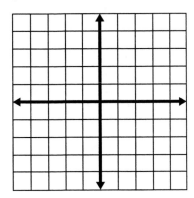

2. $y = (2/3)x + 1$

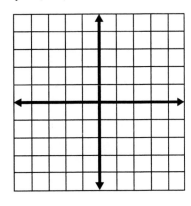

3. $4x + 2y = -6$

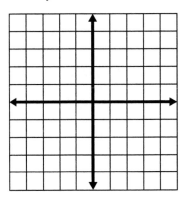

4. $3x - 2y = 2$

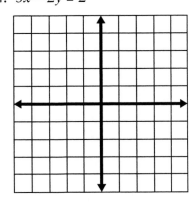

5. $x + 2y = 4$

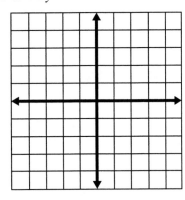

6. $3x - 2y = -4$

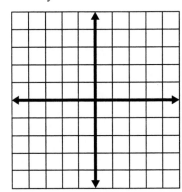

7. $4x + 4y = 0$

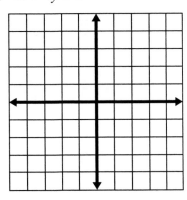

8. $2x + 4y = -12$

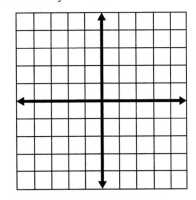

The Complete Book of Graphing

Name: _____

Date: _____ Period: _____

GRAPHING SYSTEMS OF EQUATIONS

To solve a system of equations by graphing, **both equations** must be graphed. The solution of the system is the point at which the **two lines intersect**.

EXAMPLE: Solve the following system of equations:

$$y = 2x - 1$$

$$y = (1/2)x + 2$$

Solution: (2,3)

> **Remember:** The solution must be verified by checking it in **both** equations.

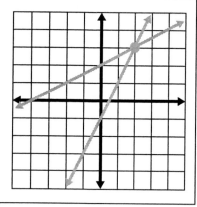

DIRECTIONS: Solve the following systems of equations:

1. Equations:

$$y = 3x - 2$$

$$y = x + 2$$

Solution: (_____)

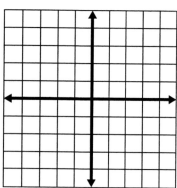

2. Equations:

$$y = x - 5$$

$$2x + 3y = 0$$

Solution: (_____)

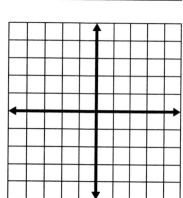

Name: _____

Date: _____ Period: _____

IDENTIFYING AND GRAPHING PARALLEL AND PERPENDICULAR LINES

Parallel lines are lines that lie in the same plane (like a desktop) but never touch, thus they have no solution. The key to identifying parallel lines is that they have the same slope, but different *y*-intercepts.

Perpendicular lines intersect at right angles. These lines have slopes that are negative inverses of each other. Perpendicular lines have a single point as a solution.

EXAMPLE 1

PARALLEL LINES $y = 2x - 3$
$y = 2x + 1$

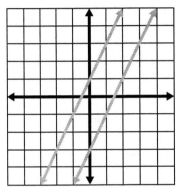

No solution—No intersection

EXAMPLE 2

PERPENDICULAR LINES $y = 2x - 1$
$y = (-1/2)x + 2$

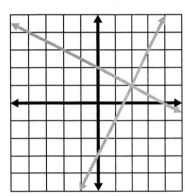

Solution: (2,1)

DIRECTIONS: Graph the following parallel or perpendicular lines and find any possible solutions:

1. $y = (3/2)x - 3$

$y = (-2/3)x + (4/3)$

Solution: _____

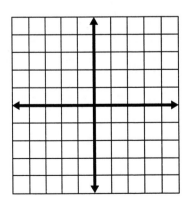

2. $4x + 2y = 8$

$y = -2x - 3$

Solution: _____

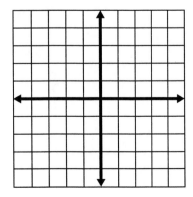

49

The Complete Book of Graphing

Name: _____

Date: _____ Period: _____

MIXED REVIEW OF GRAPHING LINEAR EQUATIONS

DIRECTIONS: Graph the following linear equations and find solutions for systems of equations:

1. $y = -3x + 2$

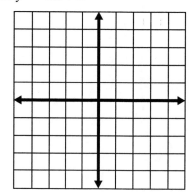

2. $y = (3/2)x - 1$

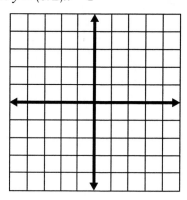

3. $3x + 3y = -6$

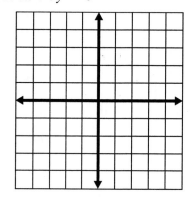

4. $4x - 2y = 0$

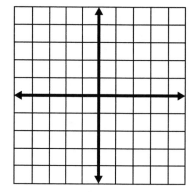

5. $y = -2x + 2$

$y = x + 5$

Solution: _____

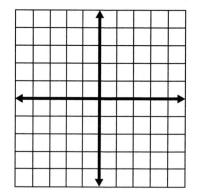

6. $x + 2y = 6$

$y = (-1/2)x - 3$

Solution: _____

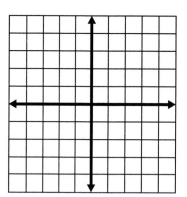

Name: _____

Date: _____ Period: _____

GRAPHING LINEAR INEQUALITIES

Graphing a linear inequality is identical to graphing a normal linear equality, except for one thing: One side of the graph will be shaded to reflect the area that has points that will make the inequality true. For example, "$y \leq 2x - 1$" is graphed as follows:

The **dark line** is the graph of the equation. The **lighter lines** indicate which side of the equation is shaded to make the inequality true. To check which side of the line is shaded, pick a point on both sides of the line and see if it makes a true statement.

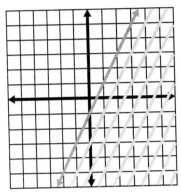

For example, using point $(0,0)$, the inequality would read $0 \leq 2(0) - 1$, which is false. Thus this side is not shaded. The point $(2,1)$ would read $1 \leq 2(2) - 1$, which is true. This side would be shaded. In this inequality, every point on the right side will be true.

DIRECTIONS: Graph and shade the following inequalities:

1. $y \leq -2x + 3$

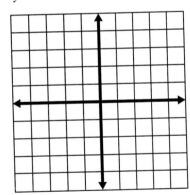

2. $y \geq 3x - 4$

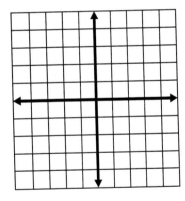

The Complete Book of Graphing

Name: _____

Date: _____ Period: _____

GRAPHING LINEAR INEQUALITIES

Graphing linear inequalities uses all inequality symbols. If the inequality uses ≤ or ≥, then the line itself is part of the solution and is graphed like this: ◄————————►

If the inequality uses < or >, then the line is not part of the solution and is graphed like this: ◄ - - - - - - -►

This is equivalent to an open circle on a number line.

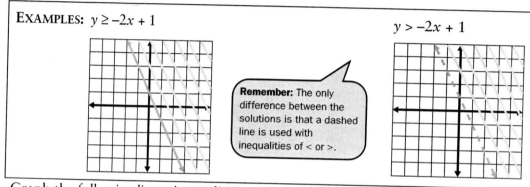

EXAMPLES: $y \geq -2x + 1$ $y > -2x + 1$

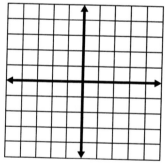

Remember: The only difference between the solutions is that a dashed line is used with inequalities of < or >.

DIRECTIONS: Graph the following linear inequalities:

1. $y \geq 2x - 3$

2. $y < x - 2$

3. $y > (2/3)x$

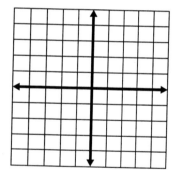

4. $y \leq -3x + 2$

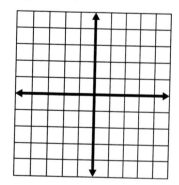

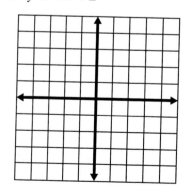

The Complete Book of Graphing

Name: _____

Date: _____ Period: _____

GRAPHING LINEAR INEQUALITIES

DIRECTIONS: Graph the following linear inequalities:

1. $y > -x + 3$

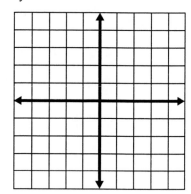

2. $y \leq (1/4)x - 4$

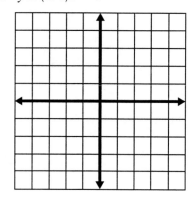

3. $y < 2x - 4$

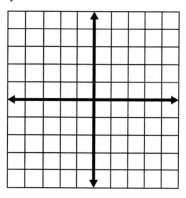

4. $y \geq -3x + 5$

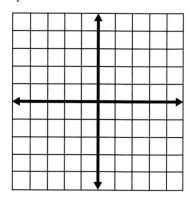

5. $y \leq x$

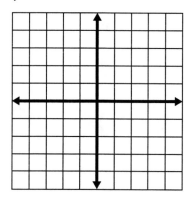

6. $y \geq (-2/3)x - 1$

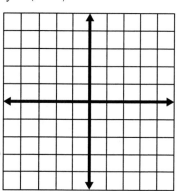

7. $y > 2x - 1$

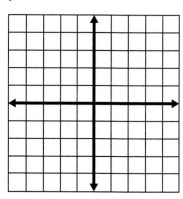

8. $y > -4x + 3$

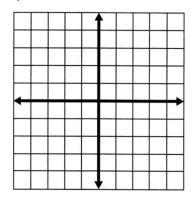

The Complete Book of Graphing

Name: _____

Date: _____ Period: _____

GRAPHING SYSTEMS OF LINEAR INEQUALITIES

Graphing systems of linear inequalities is the same as graphing one inequality, except that there are two shaded areas.

EXAMPLE: $y > 2x - 1$ and $y < (-1/3)x + 2$

Remember: Be sure to use the dashed line for < or >. Also, the system **can** contain both dashed and whole lines in the solution.

The solution is the area where the lines cross.

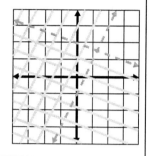

DIRECTIONS: Graph the following systems of linear inequalities:

1. $y > -2x + 1$

   $y < x - 2$

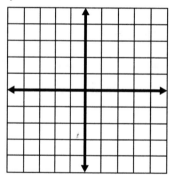

2. $y \leq (3/4)x - 2$

   $y \geq -x + 3$

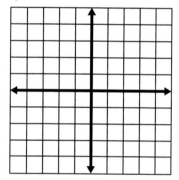

3. $y < x + 1$

   $y > -x - 1$

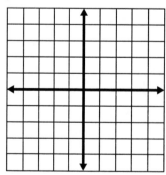

4. $y \geq (-1/2)x$

   $y < -2x + 3$

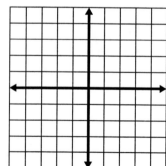

Name: _____

Date: _____ Period: _____

GRAPHING DATA POINTS AND FINDING THE SLOPE

Graphing points from data is an effective method of seeing relationships between two variables. Finding the slope is needed to understand what the data describe.

EXAMPLE: The following measurements are the heights of ten students in inches and then centimeters. Using the data, find the slope and determine what the slope means.

1. (60; 152.4)
2. (66; 167.64)
3. (73; 185.42)
4. (62; 157.48)
5. (72; 182.88)
6. (68; 172.72)
7. (70; 177.8)
8. (65; 165.1)
9. (71; 180.34)
10. (63; 160.02)

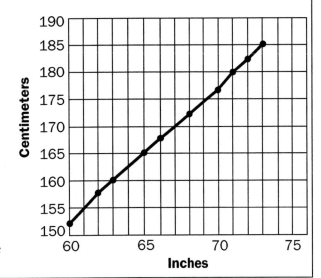

The points on this graph represent the students' heights. The slope of the line is approximately 2.54. The slope represents the conversion rate from inches to centimeters.

DIRECTIONS: The following measurements are the weights of ten students in kilograms and in pounds. Graph the relationship between the two measurements. Find the slope, then determine what the slope means.

1. (50; 110)
2. (55; 121)
3. (60; 132)
4. (65; 143)
5. (70; 154)
6. (75; 165)
7. (80; 176)
8. (85; 187)
9. (90; 198)
10. (95; 209)

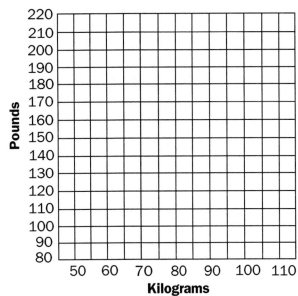

Slope: _____

Meaning: _____

_____

Name: _____

Date: _____ Period: _____

GRAPHING DATA POINTS AND FINDING THE SLOPE

1. Andrea is starting a new workout schedule and has bought a bike so she can get out of the gym and get some fresh air. She has read that the best exercising speed on a bicycle is about 8 miles per hour. The following data points are for a person who is moving at that speed. The slope of the line that is plotted will be the person's speed in ft/sec. The reason this is possible is because speed (V), distance (D), and time (T) are all connected by the formula:

$$D = V \times T$$

D is measured in feet, V is measured in ft/sec, and T is measured in seconds. Graph each point and determine the slope of the line using any two points. At 8 miles per hour, what will Andrea's velocity be?

Time (sec)	Distance (ft)
1	12
2	24
3	36
4	48
5	60
6	72
7	84

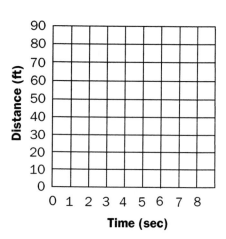

Velocity = _____ ft/sec

2. As Ethan is riding on his skateboard, he notices that every time he pushes with his foot he not only keeps moving but also continually speeds up. This process is called acceleration and is part of the formula:

$$F = m \times a$$

As he pushes, the force he pushes with increases and the acceleration of the skateboard increases. What doesn't change is the mass of the skateboard, and the slope of the line that compares force and acceleration will reveal the skateboard's mass in kilograms. Graph each point and determine the slope of the line using any two points.

Question: What is the mass of the skateboard in pounds? (1 kilogram is equivalent to 2.2 pounds.)

Force (newtons)	Acceleration (m/s^2)
20	5
40	10
60	15
80	20
100	25
120	30
140	35
160	40

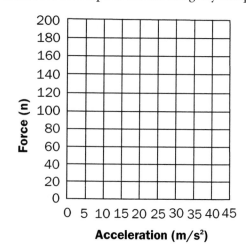

Mass = _____ lbs

(continued)

The Complete Book of Graphing

Name: _____

Date: _____ Period: _____

GRAPHING DATA POINTS AND FINDING THE SLOPE *(continued)*

3. In a car going 55 miles an hour, you can cover a large number of feet every second. Graph each point and determine the slope of the line using any two points. The slope of this line will reveal the speed of that car in feet/second.

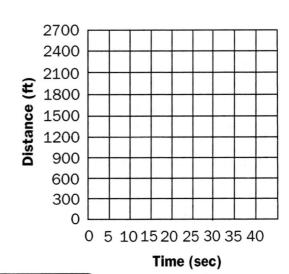

Time (sec)	Distance (ft)
0	0
5	400
10	800
15	1200
20	1600
25	2000
30	2400

Velocity = _____

Remember: Include units in the final answer.

4. In question 2 you found the mass of a skateboard using the slope of a line you graphed. With the following data points, find the mass of the object that was accelerated. Use the same method of graphing and calculating the slope from two data points. Remember that force = mass × acceleration ($F = m \times a$).

Question: Could this object be a person?

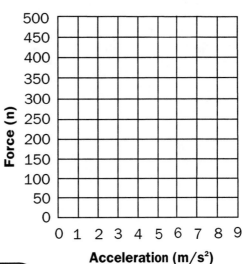

Force (newtons)	Acceleration (m/s^2)
0	0
45	1
90	2
135	3
225	5
315	7
360	8
405	9

Mass = _____

Remember: Include units in the final answer.

© 2001 J. Weston Walch, Publisher

The Complete Book of Graphing

Name: _____

Date: _____ Period: _____

GRAPHING QUADRATIC EQUATIONS

The quadratic equation has the form $y = ax^2 + bx + c$. The equation forms a curve called a **parabola** when graphed. A table helps to demonstrate how the variables are related.

EXAMPLE: $y = x^2$

Table

x	y	Coordinates
3	9	(3,9)
2	4	(2,4)
1	1	(1,1)
0	0	(0,0)
−1	1	(−1,1)
−2	4	(−2,4)
−3	9	(−3,9)

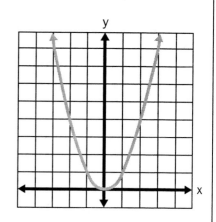

The bottom of the graph is called the **vertex** (called vertices if more than one). The vertical line (y-axis in this graph) is called the **axis**. However, not every parabola has (0,0) for a vertex. The x value of the vertex will always be halfway between the x-intercepts. Use this value to find the y value for the vertex. To find these intercepts, solve for where $y = 0$.

Example: $y = x^2 - 2x - 3$

If $y = 0$, then the equation is $x^2 - 2x - 3 = 0$. This becomes $(x - 3)(x + 1) = 0$, and when x is solved, it reveals x as −1 and 3. Thus, the x-intercepts are (3,0) and (−1,0).

To find the y-intercept, simply set $x = 0$. For this example, $y = 0^2 - 2(0) - 3$, which reveals the y-intercept as (0,−3).

Remember: Not every parabola faces up, and not every parabola has two x-intercepts. Some have only **one** or **none,** so use tables when needed.

DIRECTIONS: Find the x-intercepts, y-intercepts, and vertices of the following quadratic equations:

Equation	x-intercept(s)	y-intercept	Vertex
1. $y = x^2 - 3x - 4$	_____	_____	_____
2. $y = x^2 - 4$	_____	_____	_____
3. $y = -2x + 1$	_____	_____	_____

The Complete Book of Graphing

Name: _____

Date: _____ Period: _____

GRAPHING QUADRATIC EQUATIONS

DIRECTIONS: Find the *x*-intercepts, *y*-intercepts, and vertices of the following quadratic equations:

Equation	*x*-intercept(s)	*y*-intercept	Vertex
1. $y = 2x^2$	_____	_____	_____
2. $y = x^2 + 2x - 3$	_____	_____	_____
3. $y = -x^2 - 9$	_____	_____	_____
4. $y = x^2 - 2x + 1$	_____	_____	_____
5. $y = -x^2 + 4$	_____	_____	_____

DIRECTIONS: Graph the following quadratic equations:

1. $y = 2x^2$

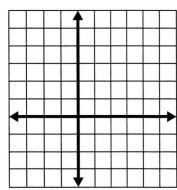

2. $y = x^2 - 1$

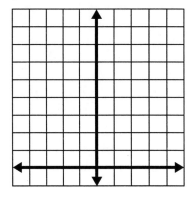

3. $y = x^2 - 4x + 3$

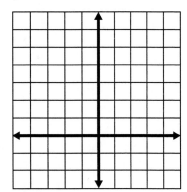

4. $y = -x^2 + 9$

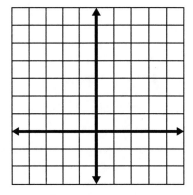

The Complete Book of Graphing

Name: _____

Date: _____ Period: _____

GRAPHING QUADRATIC EQUATIONS

Using the following quadratic equations, find the *x*-intercepts, *y*-intercepts, and vertices and then complete the graph of each equation:

1. $y = x^2 + 1$

   x-intercept(s): _____

   y-intercept: _____

   Vertex: _____

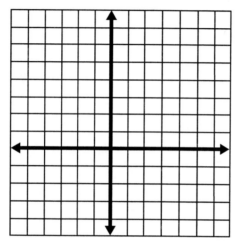

2. $y = -2x^2 + 2$

   x-intercept(s): _____

   y-intercept: _____

   Vertex: _____

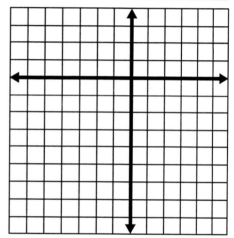

3. $y = x^2 - 2x + 1$

   x-intercept(s): _____

   y-intercept: _____

   Vertex: _____

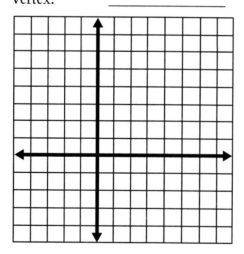

4. $y = x^2 + 3x - 4$

   x-intercept(s): _____

   y-intercept: _____

   Vertex: _____

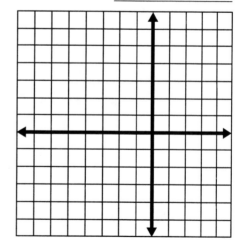

Name: _____

Date: _____ Period: _____

GRAPHING CIRCLES

The standard form of a circle is

$$(x - h)^2 + (y - k)^2 = r^2.$$

The r is the radius of the circle (*not r^2*). The h and k terms give the **center** of the circle. The graph of a circle looks as follows:

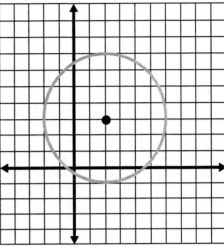

EXAMPLE:	$(x - 2)^2 + (y - 3)^2 = 16$
	This is $(x - 2)^2 + (y - 3)^2 = 4^2$ in standard form. The radius is 4, and the center is (2,3).

DIRECTIONS: Find the radius and center of each of the following circular equations, then graph the circle:

1. $(x - 1)^2 + (y - 2)^2 = 9$

    Radius: _____ Center: (_____)

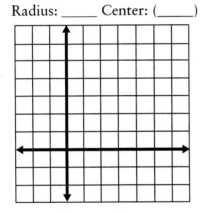

2. $x^2 + y^2 = 25$

    Radius: _____ Center: (_____)

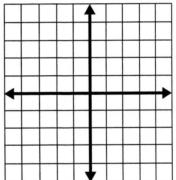

3. $(x + 2)^2 + (y + 1)^2 = 4$

    Radius: _____ Center: (_____)

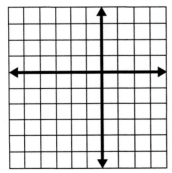

4. $(x - 1)^2 + (y + 3)^2 = 16$

    Radius: _____ Center: (_____)

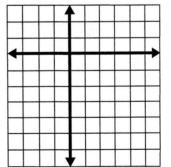

The Complete Book of Graphing

Name: _____

Date: _____ Period: _____

GRAPHING CIRCLES

DIRECTIONS: Find the radius and center of each of the following circular equations, then graph the circle:

1. $(x - 1)^2 + (y - 2)^2 = 16$
    Radius: ____ Center: (____)

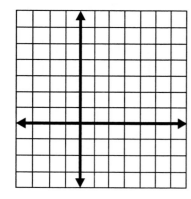

2. $x^2 + (y - 4)^2 = 9$
    Radius: ____ Center: (____)

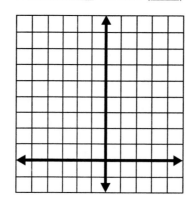

3. $(x + 2)^2 + (y - 3)^2 = 4$
    Radius: ____ Center: (____)

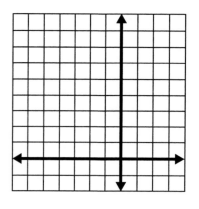

4. $(x + 1)^2 + (y + 1)^2 = 1$
    Radius: ____ Center: (____)

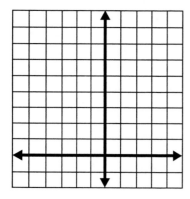

5. $(x - 4)^2 + (y + 2)^2 = 4$
    Radius: ____ Center: (____)

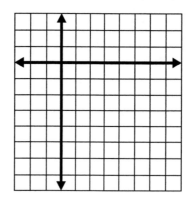

6. $(x + 4)^2 + y^2 = 9$
    Radius: ____ Center: (____)

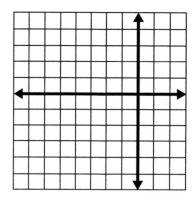

Name: _____

Date: _____ Period: _____

GRAPHING CIRCLES

DIRECTIONS: Find the radius and center of each of the following circular equations, then graph the circle:

1. $(x - 3)^2 + (y - 1)^2 = 9$

Radius: _____ Center: (_____)

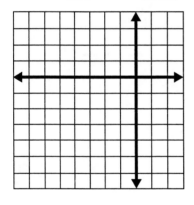

2. $x^2 + y^2 = 16$

Radius: _____ Center: (_____)

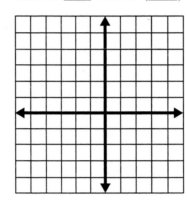

3. $(x - 2)^2 + (y + 2)^2 = 4$

Radius: _____ Center: (_____)

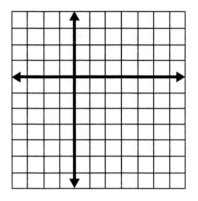

4. $(x + 3)^2 + (y + 2)^2 = 16$

Radius: _____ Center: (_____)

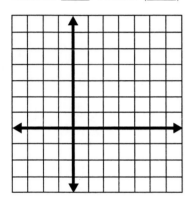

5. $(x - 1)^2 + (y - 1)^2 = 25$

Radius: _____ Center: (_____)

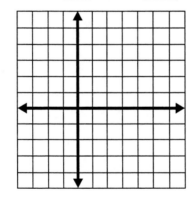

6. $x^2 + (y - 3)^2 = 1$

Radius: _____ Center: (_____)

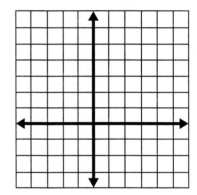

The Complete Book of Graphing

Name: _____

Date: _____ Period: _____

GRAPHING ELLIPSES

The standard form of an ellipse is

$$\frac{x^2}{a^2} + \frac{y^2}{b^2} = 1$$

In this case, the radius is not constant, as in the circle. It has intersection points, which are derived from the a and b terms.

EXAMPLE:	$\dfrac{x^2}{36} + \dfrac{y^2}{16} = 1$

The points of intersection would be:

(6,0) (–6,0) (0,4) (0,–4)

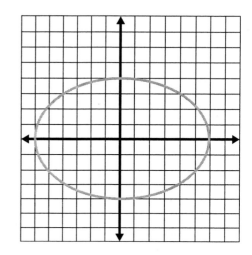

DIRECTIONS: Find the points of intersection of each of the following equations, then graph the ellipse:

1. $\dfrac{x^2}{16} + \dfrac{y^2}{9} = 1$ Points

(_____)

(_____)

(_____)

(_____)

2. $\dfrac{x^2}{25} + \dfrac{y^2}{4} = 1$ Points

(_____)

(_____)

(_____)

(_____)

3. $\dfrac{x^2}{4} + \dfrac{y^2}{9} = 1$ Points

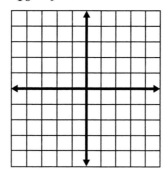

(_____)

(_____)

(_____)

(_____)

4. $\dfrac{x^2}{25} + \dfrac{y^2}{16} = 1$ Points

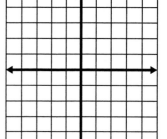

(_____)

(_____)

(_____)

(_____)

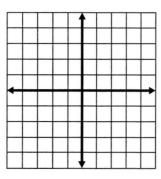

Name: _____

Date: _____ Period: _____

GRAPHING ELLIPSES

To graph an ellipse whose center is not at the origin, the formula used is:

$$\frac{(x-h)^2}{a^2} + \frac{(y-k)^2}{b^2} = 1$$

The center of this ellipse is (h,k).

This resembles the formula of a circle.

EXAMPLE:	$\dfrac{(x-2)^2}{36} + \dfrac{(y-1)^2}{25} = 1$

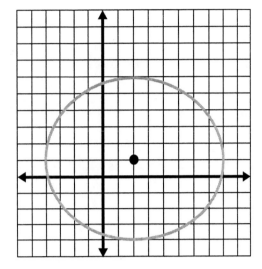

Center: (2,1)

Horizontal vertices: (–4,1), (8,1)

Vertical vertices: (2,–4), (2,6)

The vertices do not intersect with the axes and are based on the center and distances of the terms a and b.

DIRECTIONS: Use the following equations to find the center and vertices; then graph each ellipse:

1. $\dfrac{(x-3)^2}{9} + \dfrac{(y-1)^2}{25} = 1$

Center: _____

Horizontal vertices: _____

Vertical vertices: _____

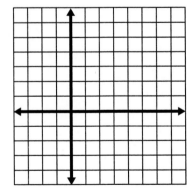

2. $\dfrac{(x+2)^2}{4} + \dfrac{(y-2)^2}{16} = 1$

Center: _____

Horizontal vertices: _____

Vertical vertices: _____

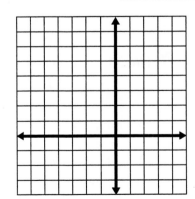

The Complete Book of Graphing

Name: _____

Date: _____ Period: _____

GRAPHING ELLIPSES

In graphing ellipses, the pieces of information most commonly needed are the **center, vertices, foci, major axis,** and **minor axis.** Using this ellipse,

$$\frac{x^2}{25} + \frac{y^2}{9} = 1$$

Center: (0,0)

Vertices: (5,0), (−5,0), (0,3), (0,−3).

The major axis is $2a$ and the minor axis is $2b$.

Major axis = 10

Minor axis = 6

The foci (c) are found from the equation $c^2 = a^2 - b^2$. In this example, $c^2 = 25 - 9$, thus $c = 4$.

From this, the foci would be (4,0) and (−4,0), using the center, and will lie on the major axis.

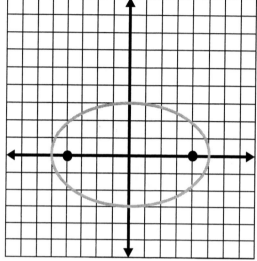

Remember: a^2 will be the larger of the denominators, so you will not have a negative radical for c.

DIRECTIONS: Use the following equations to find the center, vertices, foci, and major and minor axes of each ellipse:

1. $\dfrac{x^2}{9} + \dfrac{y^2}{25} = 1$

Center: _____

Horizontal vertices: _____

Vertical vertices: _____

Major axis: _____

Minor axis: _____

Foci: _____

2. $\dfrac{(x+3)^2}{100} + \dfrac{(y-1)^2}{64} = 1$

Center: _____

Horizontal vertices: _____

Vertical vertices: _____

Major axis: _____

Minor axis: _____

Foci: _____

The Complete Book of Graphing

Name: _____

Date: _____ Period: _____

GRAPHING HYPERBOLAS

Hyperbolas use three general equations for graphing:

1. $\dfrac{(x-h)^2}{a^2} - \dfrac{(y-k)^2}{b^2} = 1$ The center is (h,k) and the foci are $(h + c,k)$ and $(h - c,k)$, where $c^2 = a^2 + b^2$.

2. $\dfrac{(y-k)^2}{a^2} - \dfrac{(x-h)^2}{b^2} = 1$ The center is (h,k) and the foci are $(h,k + c)$ and $(h,k - c)$, where $c^2 = a^2 + b^2$.

3. $xy = k$ The center is the origin, and the graphs lie in Quadrants I and III if $k > 0$ and in Quadrants II and IV if $k < 0$.

The first graph opens left and right, the second opens up and down, the third opens diagonally—if k is positive, it opens down to the left and up to the right and the opposite occurs if k is negative.

The general appearances of the graphs are as follows:

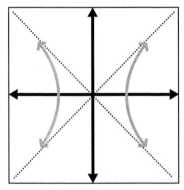 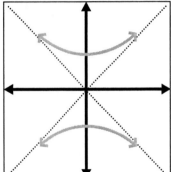 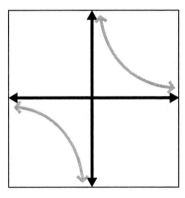

DIRECTIONS: Find the needed information using the following equations:

1. $\dfrac{(x-2)^2}{9} - \dfrac{(y-3)^2}{16} = 1$

     Center: _____

     Foci: _____

     Opens how? _____

2. $xy = 4$

     Center: _____

     Opens how? _____

3. $\dfrac{(y+3)^2}{1} - \dfrac{(x-1)^2}{4} = 1$

     Center: _____

     Foci: _____

     Opens how? _____

The Complete Book of Graphing

Name: _____

Date: _____ Period: _____

GRAPHING HYPERBOLAS

DIRECTIONS: Find the following information and graph each equation:

> **Remember:** The vertices are the points of symmetry for the graph.

1. $\dfrac{x^2}{4} - \dfrac{y^2}{25} = 1$

   Center: _____
   Vertices: _____
   Foci: _____

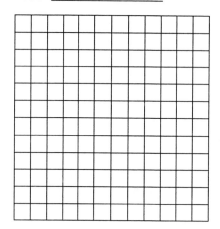

2. $\dfrac{(y-2)^2}{36} - \dfrac{(x+1)^2}{64} = 1$

   Center: _____
   Vertices: _____
   Foci: _____

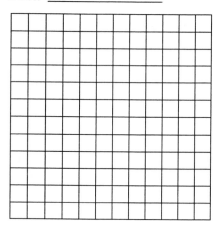

3. $xy = -25$

   Center: _____
   Vertices: _____

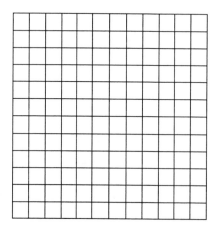

4. $\dfrac{(x+1)^2}{2} - \dfrac{(y+3)^2}{2} = 1$

   Center: _____
   Vertices: _____
   Foci: _____

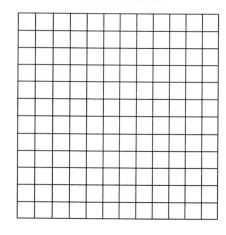

Name: _____

Date: _____ Period: _____

GRAPHING HYPERBOLAS

Remember: The vertices are the points of symmetry for the graph.

DIRECTIONS: Find the following information and graph each equation:

1. $\dfrac{x^2}{4} - \dfrac{y^2}{5} = 1$

Center: _____
Vertices: _____
Foci: _____

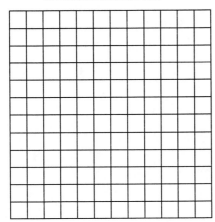

2. $\dfrac{(y-2)^2}{9} - \dfrac{(x+1)^2}{16} = 1$

Center: _____
Vertices: _____
Foci: _____

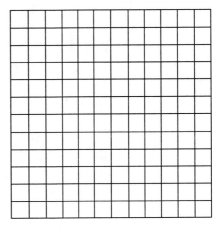

3. $xy = 20$

Center: _____
Vertices: _____

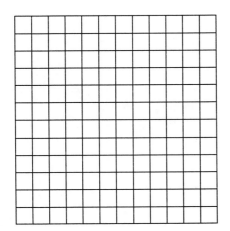

4. $\dfrac{(x-4)^2}{3} - \dfrac{(y+1)^2}{6} = 1$

Center: _____
Vertices: _____
Foci: _____

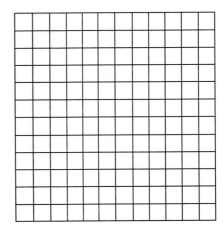

The Complete Book of Graphing

Name: _____

Date: _____ Period: _____

PRACTICAL APPLICATIONS

1. A trucker must make a delivery 400 miles away by the end of the day. Graph the driver's possible average rate of speed compared to travel time, and find his minimum average speed if his day begins at 9:00 am and ends at 5:00 pm. Use the equation $xy = 400$.

2. Concentric circles have the same center but different radii. Write the set of equations to model a dart board with concentric circles and diameters of 6 cm, 12 cm, 18 cm, 24 cm, and 30 cm.

3. An airplane drops supplies to firefighters at a low altitude. The supplies reach the ground in 10 seconds. Using the equation $s = 1/2\ gt^2$, with s = distance, g = gravity (9.8 m/s^2), and t = time, find the approximate altitude from which the plane dropped the supplies.

4. A ball dropped from the Royal Gorge Bridge in Colorado (approximately 315 meters tall) will take about how long to hit the bottom of the gorge? Use the formula $s = 1/2gt^2$, where s = distance, g = gravity (9.8 m/s^2), and t = time.

5. A rocket is shot straight into the air with an initial velocity of 96 feet/second. The formula $h(t) = 96t - 16t^2$ shows its height after t seconds. Find the maximum height of the rocket and the time it will hit the ground.

Name: _____

Date: _____ Period: _____

PRACTICAL APPLICATIONS

1. A ball is dropped from a 1000-meter tower. How long will it take to reach the ground? Use the equation $s = 1/2gt^2$, where s = distance, g = gravity (9.8 m/s^2), and t = time.

2. A cannon shoots a cannonball into the air with an initial velocity of 240 feet per second. Using the formula $h(t) = 240t - 16t^2$, what is the maximum height of the cannonball and how long will it be before it hits the ground?

3. An archer shoots an arrow into the air from the edge of a cliff 240 feet above sea level with a velocity of 32 feet per second. Using the formula $h(t) = 240 + 32t - 16t^2$, find the time it takes the arrow to fall back to sea level.

4. A manufacturer of wheels has daily production costs of $C(x) = 600 - 25x + 0.25x^2$, where C is the total cost in dollars and x is the number of units produced. How many wheels should be produced daily to minimize the cost per wheel?

5. The revenue for a company is shown by the equation $R = 100x - 0.0025x^2$, where R is revenue and x is the number of units produced. Find the value of x that will maximize the revenue.

Name: _____

Date: _____ Period: _____

MIXED REVIEW OF GRAPHS

DIRECTIONS: Graph each of the following equations:

1. $(x - 2)^2 + (y - 2)^2 = 9$

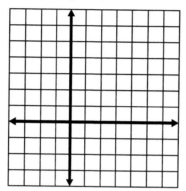

2. $x^2 + y^2 = 25$

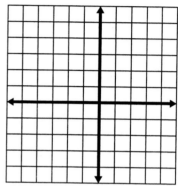

3. $3x + 6y = -12$

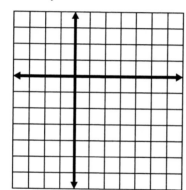

4. $y = (1/4)x - 2$

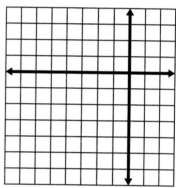

5. $y = x^2 + 1$

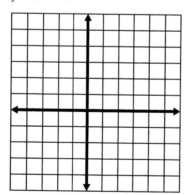

6. $y = (x - 2)^2 - 2$

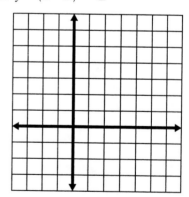

The Complete Book of Graphing

Name: _____

Date: _____ Period: _____

MIXED REVIEW OF GRAPHS

DIRECTIONS: Graph the following equations:

1. $y = -2x + 3$

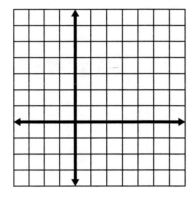

2. $y = -(x + 1)^2 + 3$

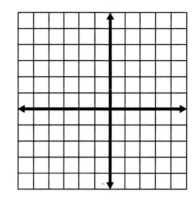

3. $(x - 2)^2 + (y + 2)^2 = 16$

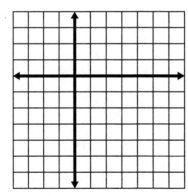

4. $xy = 5$

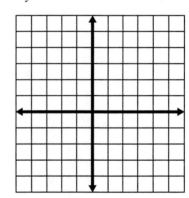

5. $\dfrac{x^2}{16} + \dfrac{y^2}{25} = 1$

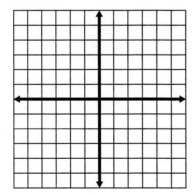

6. $\dfrac{(x - 1)^2}{5} - \dfrac{(y - 2)^2}{1} = 1$

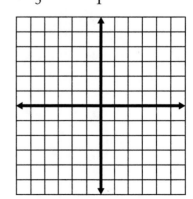

Name: _____

Date: _____ Period: _____

BAR GRAPHS

Bar graphs are one of the most common forms of graph. They can be **horizontal** or **vertical,** and are uniform in spacing to accurately compare various conditions of a specific quantity. The following is an example using the wind chill at 50°F.

Wind Speed (mph)	Equivalent Temperature (°F)
0	50
5	48
10	40
15	32
20	30
25	28

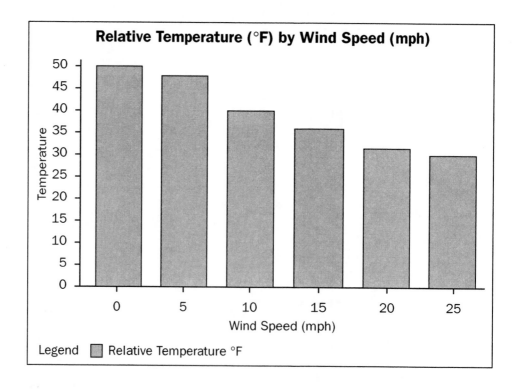

(continued)

Name: _____

Date: _____ Period: _____

BAR GRAPHS *(continued)*

DIRECTIONS: Construct a bar graph to effectively depict each set of data.

> **Remember:** Always label each axis and use equal spacing.

1. In a class of 25 students, here are the grades on a math test:

Number of students	Grade
4	A
7	B
9	C
4	D
1	F

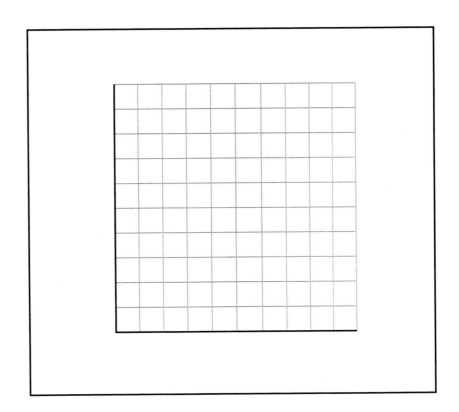

(continued)

Name: _____

Date: _____ Period: _____

BAR GRAPHS *(continued)*

2. In some parts of the country, the amount of rain that falls varies widely, depending on the time of year. The figures here give the rainfall in Anywhere, USA. Graph these values to see if the rainfall shows any kind of trend. *Questions:* During which season is the rainfall heaviest? _____ In which season is it lightest? _____

Month	Rainfall (in.)
Jan.	4.2
Feb.	3.5
Mar.	3.2
Apr.	2.0
May	1.3
June	1.1
July	0.9
Aug.	1.4
Sept.	1.8
Oct.	2.3
Nov.	3.1
Dec.	3.6

(continued)

Name: _____

Date: _____ Period: _____

BAR GRAPHS *(continued)*

3. pH is a measurement used by scientists to determine how "strong" certain chemicals and substances are in their ability to react with the things around them. The pH scale goes from 0 to 14, where 7 is neutral, 0 is a very strong acid (like sulfuric acid), and 14 is a very strong base (like concentrated drain cleaner). Graph the pH for each of the susbtances. *Questions:* Which three substances are closest in value? Why might that be?

Substance	pH
Pure water	7.0
Blood	7.6
Black coffee	5.0
Lemon juice	2.4
Sea water	8.2
Ammonia	11.4
Bleach	12.5

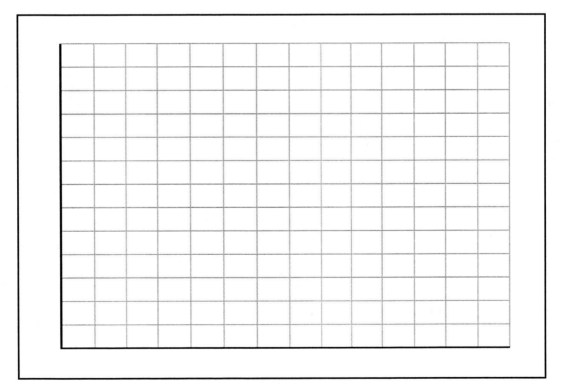

Name: _____

Date: _____ Period: _____

HISTOGRAMS

Histograms are similar to bar graphs, but with histograms, the bars always touch and represent a **quantitative value**. The histogram divides each "bar" by **class** with **upper** and **lower limits** and **midpoints**. This is compared with the frequency with which values lie within the limits of each class. An example would be the viewer ages of the movie *Star Fighters*. The ages range from 5 to 65. The results are as follows:

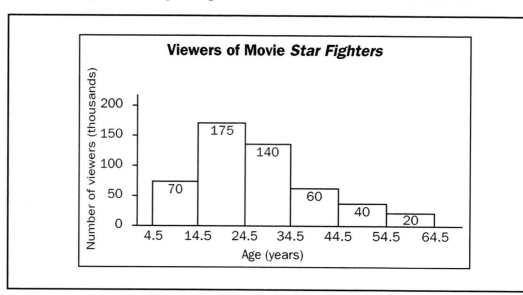

This histogram shows that members of the largest group of viewers were between the ages of 15 and 24. Each class must be separate and cannot overlap. In this graph, the ages cannot be listed at intervals of 5, 15, 25, 35, 45, 55, and 65 because a viewer at these ages would be listed in two different classes, so intervals are set up to prevent this. Each class uses the frequency of data points within its class width. In the case of this histogram, there were about 170,000 viewers between the ages of 15 and 24.

To find the class width, use the following formula:

$$\frac{\text{Largest data value} - \text{smallest data value}}{\text{Desired number of classes}} \approx \text{class width} \implies \frac{64-5}{6} \approx 9.8 \text{ or } 10$$

The class width will vary with the number of classes needed.

The **midpoint** of each class is simply the average of the upper limit of the class and the lower limit of the class, or $\frac{65+5}{2}$.

(continued)

Name: _____

Date: _____ Period: _____

HISTOGRAMS *(continued)*

DIRECTIONS: Using the following data, construct histograms to accurately depict each group.

1. A new company, ConGlom Inc., is trying to decide whether or not to reimburse employees for their daily commutes. The company wants to analyze the varying distances people travel to see who qualifies for reimbursement. These are the distances employees travel:

5	12	11	27	6	18	22	13	31	20	16	9
14	18	22	30	33	11	8	14	25	21	16	11
19	14	24	28	13	8	16	2	22	31	40	23
4	26	30	22	17	15	18	8	31	12	19	7

DIRECTIONS: Find the largest and smallest values.

Largest: _____   Smallest: _____

If we want six classes, what would the class width be?

Class width: ≈_____

> **Remember:** You must tally the frequency of items in each class.

DIRECTIONS: Construct the histogram:

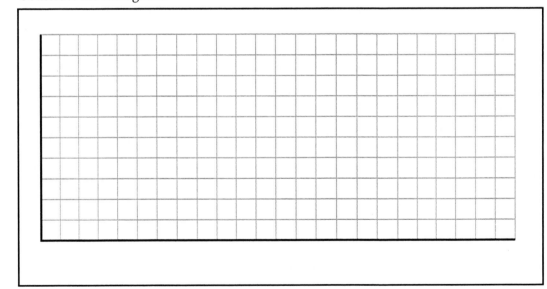

(continued)

Name: _____

Date: _____ Period: _____

HISTOGRAMS *(continued)*

2. Brian's Pancake Palace is trying to figure out why some days they run out of pancake batter, while on other days they have too much. For a month, Brian kept track of how many people ordered pancakes between 7:00 A.M. and 8:00 A.M. every day. Use six classes to show his findings on a histogram.

25	31	18	36	28	12
41	19	22	28	30	21
31	14	18	35	24	21
27	19	13	33	39	24
33	25	20	39	22	17

DIRECTIONS: Find: Largest value: _____

Smallest value: _____

Class width: ≈ _____

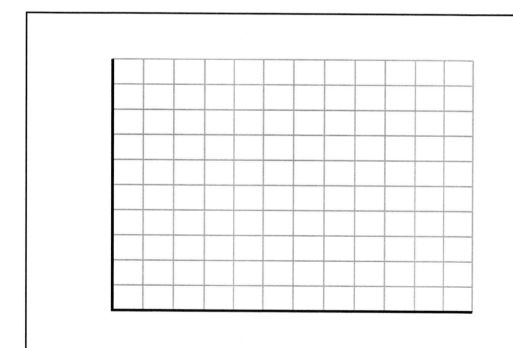

(continued)

Name: _____

Date: _____ Period: _____

HISTOGRAMS *(continued)*

3. Miki is a junior in high school. She has been getting mailings from colleges all over the country, some of which sound very interesting. However, part of her decision has to be based on cost. She has collected annual tuition costs for 40 colleges. Complete the histogram, using six classes, for the tuition costs. The numbers are in tens of thousands of dollars, so, for example, 2.1 = $21,000.

DIRECTIONS: Find: Highest: _____

Lowest: _____

Class width: ≈ _____

2.1	3.3	6.5	1.4	2.4	5.7	3.0	2.8
3.6	4.1	0.9	5.5	2.7	4.9	3.5	1.9
6.2	3.6	4.4	2.5	1.8	1.6	3.1	4.0
2.2	4.5	3.2	2.0	2.3	1.5	1.7	2.8
1.3	2.8	3.4	4.5	6.1	3.0	2.2	1.8

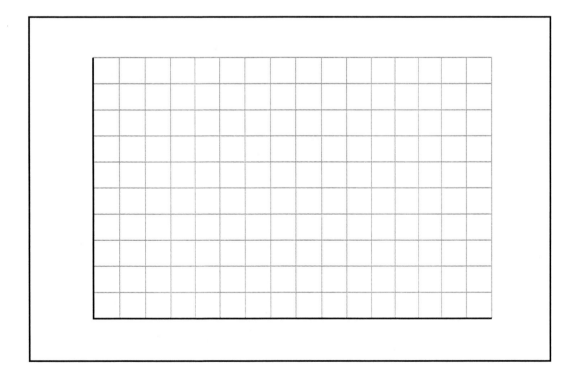

(continued)

Name: _____

Date: _____   Period: _____

HISTOGRAMS *(continued)*

4. Every year, Mr. Hutchinson evaluates his students' scores on their final exams for his class. He uses this information to make sure his tests are fair to students and cover the information in his course.

DIRECTIONS: Create a histogram for this year's grades. Use one class for each letter, A–F.

A = 92–100   D = 65–73
B = 83–91   F = 64 and below
C = 74–82

88	91	83	75	90
91	83	79	81	89
68	77	83	95	93
70	86	81	93	92
76	87	89	74	90
97	63	80	98	80
90	74	88	89	91
88	85	78	85	72

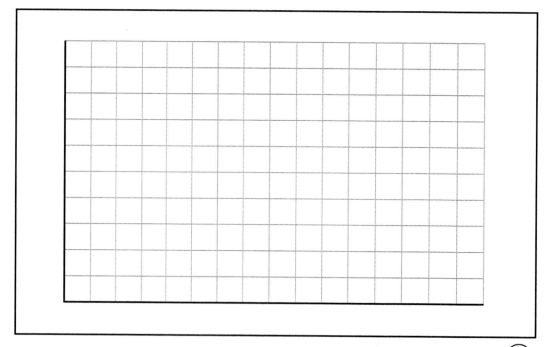

Name: _____

Date: _____ Period: _____

LINE GRAPHS

Line graphs are similar to bar graphs. They are used extensively in business and with time-oriented graphs. An example is provided using the average monthly high temperatures in Somewhere, Texas.

Jan.	Feb.	Mar.	Apr.	May	June	July	Aug.	Sept.	Oct.	Nov.	Dec.
30°F	36°F	48°F	66°F	81°F	90°F	92°F	91°F	79°F	58°F	44°F	38°F

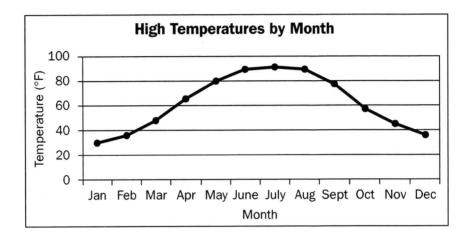

(continued)

Name: _____

Date: _____ Period: _____

LINE GRAPHS *(continued)*

DIRECTIONS: Create a line graph from the supplied data.

1. The farmers of Somewhere, Texas, are worried about frost damaging their crops. Some of the farmers claim that frost is a minor problem in such a warm state, while others say they have had crops destroyed by cold weather. Information from one year isn't enough data to make a decision, so they decided to make their plans based on the average monthly low temperatures for several years. *Question:* If frost is likely to cause the most damage when the temperature is below 32°F, what months should be avoided for growing crops?_____

_____

Jan.	Feb.	Mar.	Apr.	May	June	July	Aug.	Sept.	Oct.	Nov.	Dec.
5°F	18°F	32°F	45°F	58°F	66°F	72°F	74°F	63°F	41°F	33°F	19°F

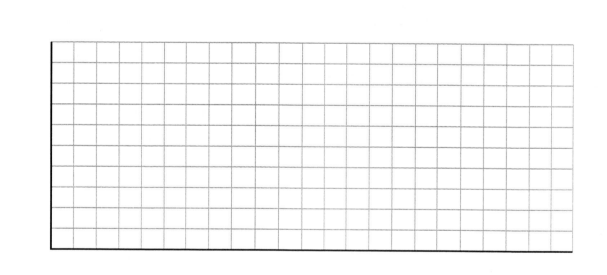

(continued)

Name: _____

Date: _____ Period: _____

LINE GRAPHS *(continued)*

DIRECTIONS: Create a line graph from the supplied data.

2. Samantha has started a computer sales business that she runs out of her home. After one year of business she is trying to see if there were any trends to people's buying habits. Graph the monthly sales data from her business. The data are in thousands of dollars, so, for example, 76 = $76,000.

Jan.	Feb.	Mar.	Apr.	May	June	July	Aug.	Sept.	Oct.	Nov.	Dec.
88	76	71	66	80	68	61	58	65	71	70	82

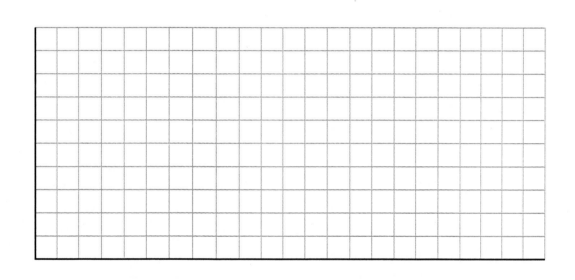

(continued)

Name: _____

Date: _____ Period: _____

LINE GRAPHS *(continued)*

DIRECTIONS: Create a line graph from the supplied data.

3. After a few years as the head of her own company, Carol has decided to see if she can cut down on employee sick days. Graph the average number of days that employees have missed per month as collected over the last three years. *Questions:* What were the four months with the most sick days? Which month doesn't seem to fit? What might be a reason for this?

Jan.	Feb.	Mar.	Apr.	May	June	July	Aug.	Sept.	Oct.	Nov.	Dec.
66	42	51	62	53	50	67	49	52	56	69	83

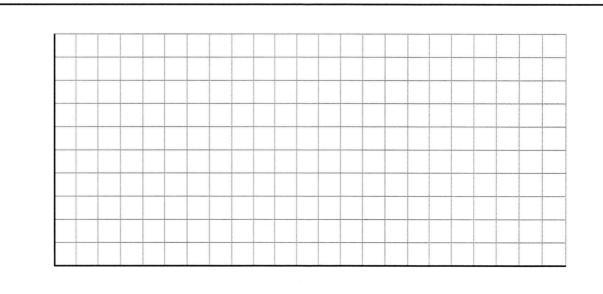

(continued)

Name: _____

Date: _____ Period: _____

PIE GRAPHS

Pie graphs are used as a method of comparing parts within a whole structure. An example would be the political affiliation of people within a city.

Somewhere, Missouri—Adult Population 14,400

Democratic	5500	Independent	2900
Republican	5200	Nonaffiliated	800

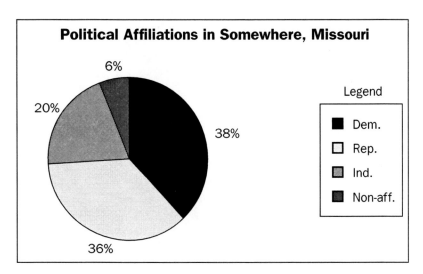

Political Affiliations in Somewhere, Missouri

Legend
- ■ Dem.
- □ Rep.
- ▨ Ind.
- ◼ Non-aff.

1. A school district is trying to address the cultural needs of all its students, but the district hasn't recently collected any information about the ethnic background of students and parents in the district. An independent research company has supplied the information, but it is not in a format that is easily presented. Put this information into a pie graph so that the information can be more easily understood.

Caucasian	11,500
Asian American	1800
Native American	3300
African American	8800
Latino	4600

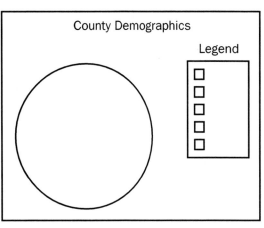

County Demographics

Legend

Remember: Always have a legend with a pie graph.

(continued)

Name: _____

Date: _____ Period: _____

PIE GRAPHS *(continued)*

2. Students in a college chemistry class collected samples of air right outside their classroom. Put the information they found into the pie graph form.

Nitrogen	78.08%
Oxygen	20.95%
Argon	0.93%
Other gases	0.04%

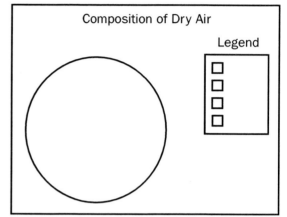

Composition of Dry Air

Legend

3. After getting his first job, Carl wants to better understand where his salary seems to disappear each month. Put his information into the pie graph form. *Question:* What percentage of his income does Carl spend on his car each month?

Mortgage	$850
Power/water	$150
Car payment	$300
Food	$150
Savings	$250
Other	$300

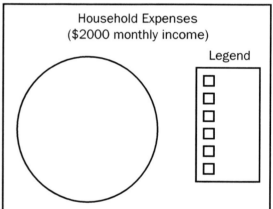

Household Expenses
($2000 monthly income)

Legend

4. People use an enormous amount of water each day. Most people use between 300 and 400 liters (77–103 gallons) per day. To get a better understanding of where that water goes, Melissa and her family measured how much water they used for their daily activities. Graph the data and give one way that Melissa and her family could reduce any one of the water-use categories she measured.

Drinking/cooking	7 L
Flushing toilets	80 L
Washing dishes	14 L
Bathing	70 L
Laundry	35 L
Miscellaneous	98 L

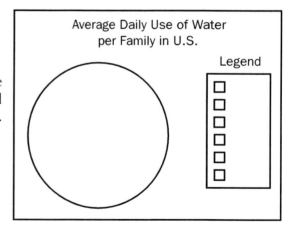

Average Daily Use of Water
per Family in U.S.

Legend

Name: _____

Date: _____ Period: _____

DUAL LINE GRAPHS

Dual line graphs compare two sets of data under the same conditions (e.g., time) to see if there is a correlation.

1. The marketing director of ConGlom Inc. wants more money for advertising. He feels sure that more advertising means more sales. Use ConGlom Inc.'s monthly sales and advertising figures to draw a dual line graph. Note: "K" means "1000," so all these figures are in thousands of dollars.

Month	Company Sales	Advertising
January	43K	2K
February	42K	1K
March	38K	1K
April	40K	3K
May	45K	3K
June	48K	3K
July	51K	4K
August	52K	4K
September	56K	4K
October	60K	5K
November	60K	5K
December	62K	5K

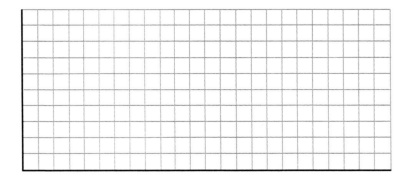

2. What does the graph indicate about advertising and sales? _____

_____

3. Would a pie graph show this relationship? _____

_____

4. What can be expected if the amount of advertising increases? _____

_____

ANSWER KEY

NUMBER LINES
PAGE 1

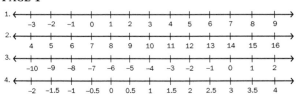

GRAPHING INEQUALITIES
PAGE 2

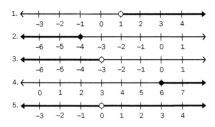

GRAPHING ABSOLUTE VALUE EQUATIONS AND INEQUALITIES
PAGE 3

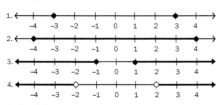

5. No solution

SOLVING AND GRAPHING INEQUALITIES
PAGE 4

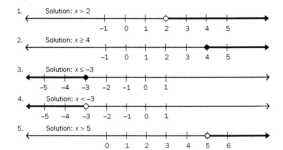

PAGE 5

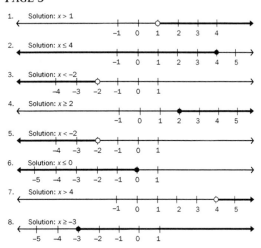

SOLVING AND GRAPHING COMBINED INEQUALITIES
PAGE 7

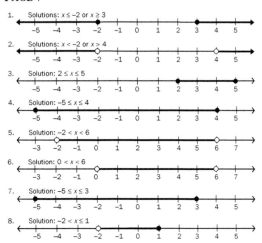

SOLVING AND GRAPHING ABSOLUTE VALUE INEQUALITIES
PAGE 8

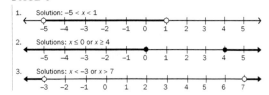

PAGE 9

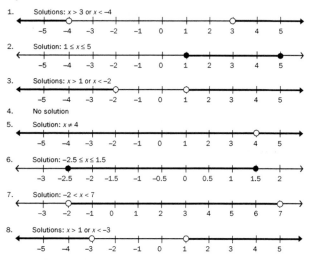

1. Solutions: $x > 3$ or $x < -4$
2. Solution: $1 \leq x \leq 5$
3. Solutions: $x > 1$ or $x < -2$
4. No solution
5. Solution: $x \neq 4$
6. Solution: $-2.5 \leq x \leq 1.5$
7. Solution: $-2 < x < 7$
8. Solutions: $x > 1$ or $x < -3$

MIXED REVIEW OF THE NUMBER LINE
PAGE 10

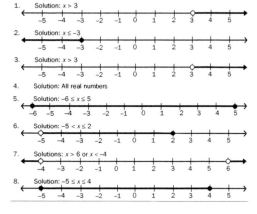

1. Solution: $x > 3$
2. Solution: $x \leq -3$
3. Solution: $x > 3$
4. Solution: All real numbers
5. Solution: $-6 \leq x \leq 5$
6. Solution: $-5 < x \leq 2$
7. Solutions: $x > 6$ or $x < -4$
8. Solution: $-5 \leq x \leq 4$

GRAPHING ORDERED PAIRS
PAGE 12

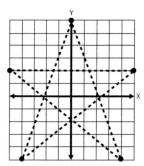

GRAPHING ORDERED PAIRS

PAGE 13

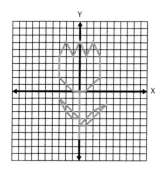

PAGE 14

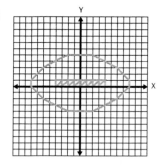

STANDARD AND SLOPE-INTERCEPT FORMS

PAGE 15

1. SF	11. SF
2. N	12. SI
3. SI	13. SI
4. SF	14. SF
5. N	15. N
6. N	16. SF
7. SF	17. N
8. SF	18. SI
9. N	19. SI
10. N	20. SF

FINDING POINTS ON A LINE

PAGE 16   Coordinates may vary.

1. (6,0) (0,2) (3,1)   3. (0,–2) (4,0) (2,–1)

2. (1,0) (0,–1) (2,1)   4. (0,–3) (1,0) (2,3)

FINDING POINTS ON A LINE (CONTINUED)

PAGE 17   Coordinates may vary.

1. (3,0) (0,–2) (6,2)
5. (–2,0) (0,4) (–1,2)

2. (3,0) (0,3) (2,1)
6. (0,–2) (6,0) (3,–1)

3. (–2,0) (0,–4) (–1,–2)
7. (0,0) (1,5) (2,10)

4. (–4,0) (0,2) (–2,1)
8. (0,8) (–4,0) (–2,4)

PAGE 18   Coordinates may vary.

1. (0,0) (3,1) (6,2)
5. (0,–3) (–3,0) (–2,–1)

2. (0,2) (2,0) (1,1)
6. (0,3) (–6,0) (2,4)

3. (0,–4) (2,0) (1,–2)
7. (0,–1) (1,0) (2,1)

4. (0,10) (–2,0) (–1,5)
8. (0,0.5) (1,–1) (–1,2)

GRAPHING LINES FROM POINTS

PAGE 19   Coordinates may vary.

1. $y = x + 3$

x	y	Coordinates
0	3	(0,3)
–3	0	(–3,0)
–1	2	(–1, 2)

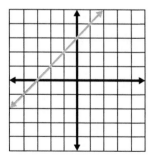

2. $y = (1/2)x - 2$

x	y	Coordinates
0	–2	(0,–2)
2	–1	(2,–1)
4	0	(4,0)

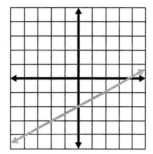

3. $y = 2x$

x	y	Coordinates
0	0	(0,0)
1	2	(1,2)
–1	–2	(–1,–2)

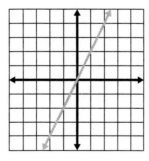

PAGE 20   Coordinates may vary.

1. $x - 2y = -4$

x	y	Coordinates
0	2	(0,2)
−4	0	(−4,0)
−2	1	(−2,1)

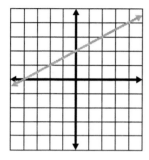

2. $y = x - 1$

x	y	Coordinates
0	−1	(0,−1)
1	0	(1,0)
2	1	(2,1)

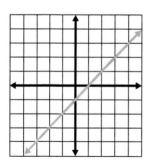

3. $y = (1/3)x + 2$

x	y	Coordinates
0	2	(0,2)
3	3	(3,3)
−3	1	(−3,1)

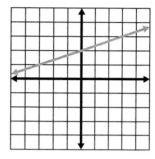

4. $2x + y = 0$

x	y	Coordinates
0	0	(0,0)
1	−2	(1,−2)
−1	2	(−1,2)

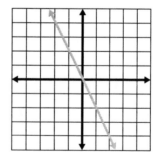

CONVERTING BETWEEN STANDARD FORM AND SLOPE-INTERCEPT FORM

PAGE 21

1. $y = (2/3)x - 3$
2. $y = (-1/2)x - 2$
3. $y = 3x - 7$
4. $y = (-5/3)x + 3$
5. $y = x - 1$
6. $y = (1/4)x - 3/2$
7. $y = (-3/2)x$
8. $y = (4/3)x - 2$
9. $y = x - 4$
10. $y = -4x + 1$

PAGE 22

1. $2x - y = 3$
2. $3x - y = -1$
3. $4x + y = 5$
4. $x + y = -2$
5. $2x - 3y = 12$
6. $x + 4y = 4$
7. $2x + y = 0$
8. $4x - 3y = 9$
9. $5x + 3y = 30$
10. $3x + y = -3$

PAGE 23

1. $m = 3 \quad b = -2$

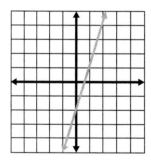

2. $m = 1/2 \quad b = 1$

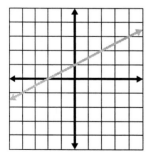

GRAPHING LINEAR EQUATIONS

PAGE 24

1. $m = 1 \quad b = 2$

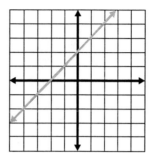

2. $m = 2 \quad b = -3$

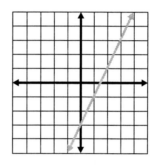

3. $m = -3 \quad b = 2$

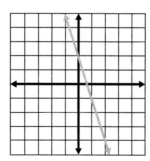

4. $m = 1/2$ $b = -1$

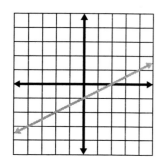

5. $m = -2/3$ $b = 2$

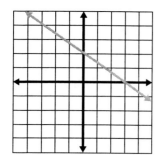

6. $m = 2$ $b = 1/2$

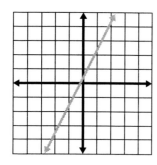

1. SI is $y = 2x - 3$.
    $m = 2$ $b = -3$

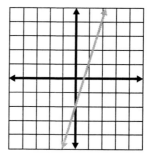

2. SI is $y = -2x + 4$.
    $m = -2$ $b = 4$

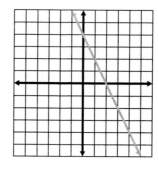

3. SI is $y = (2/3)x + 2$.
    $m = 2/3$ $b = 2$

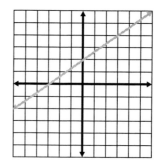

4. SI is $y = -3x$.
    $m = -3$ $b = 0$

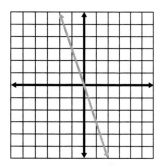

FINDING SLOPE FROM TWO POINTS

PAGE 26

1. −2	6. −1/2
2. −1/2	7. 5/3
3. 3	8. 1/3
4. 7	9. −1/2
5. −3/5	10. 0

PAGE 27

1. −1	11. −4/5
2. 2	12. 5/3
3. −9/5	13. 9/8
4. −1/8	14. −3/4
5. 7/8	15. −4/9
6. 2/5	16. −7/4
7. −1	17. −1/4
8. 5/7	18. −3/8
9. 1/2	19. 3
10. 2	20. −1/8

FINDING SLOPE FROM A POINT AND THE Y-INTERCEPT

PAGE 28

1. −1/3	9. 0
2. 3	10. −3/2
3. −5	11. −5/3
4. −3/2	12. 1
5. −2	13. 3/5
6. 1/3	14. 0
7. −2	15. −5/2
8. 6	

NO SLOPE AND ZERO SLOPE

PAGE 29

1. 0	6. 2/3
2. no slope	7. −1/2
3. −1	8. no slope
4. 0	9. 0
5. no slope	10. 0

FINDING SLOPE FROM A GRAPH

PAGE 30

1. 1	3. 1/3
2. −2	4. 0

PAGE 31

1. 1	4. 2/3
2. −1/2	5. no slope
3. 3	6. 1/2

PAGE 32

1. −1	4. 1
2. 1/3	5. −2/3
3. 4	6. 0

FINDING SLOPE FROM POINTS OR A GRAPH

PAGE 33

1. –2
2. –2/3
3. 0
4. –4/3
5. –3/10

6. –2
7. 3
8. 1/2
9. –2/3
10. 1

FINDING THE INTERCEPTS FROM A GRAPH

PAGE 34

1. x-intercept (3,0)
   y-intercept (0,–3)

2. x-intercept (–1,0)
   y-intercept (0,1)

3. x-intercept (–3,0)
   y-intercept (0,–1)

4. x-intercept (1,0)
   y-intercept (0,–3)

5. x-intercept (2,0)
   y-intercept (0,2)

6. x-intercept (–2,0)
   y-intercept (0,3)

MIXED REVIEW

PAGE 35

1.

x	y	Coordinates
0	–2	(0,–2)
4	0	(4,0)
2	–1	(2,–1)

2.

x	y	Coordinates
0	–2	(0,–2)
1	1	(1,1)
2	4	(2,4)

3. $y = (-2/3)x + 2$
4. $y = 3x + 2$
5. $2x - y = 3$
6. $3x + 4y = 4$
7. 1/4

8. 3/2
9. –3/2
10. –3
11. no slope
12. 0

FINDING THE EQUATION OF A LINE FROM THE SLOPE AND THE Y-INTERCEPT

PAGE 36

1. $y = 4x + 3$
2. $y = -3x - 1$
3. $y = x + 12$
4. $y = (1/3)x - 3$
5. $y = -2x$
6. $y = (-2/5)x + 4$

7. $y = 3$
8. $y = (4/3)x + 4/5$
9. $y = -6x + 1$
10. $y = 25x + 75$
11. $y = (-7/3)x - 6$
12. $y = 0$

FINDING THE EQUATION OF A LINE FROM THE SLOPE AND A POINT

PAGE 37

1. $y = 4x - 1$
2. $y = -2x + 8$
3. $y = 5x + 2$
4. $y = 3x + 10$
5. $y = -x - 2$

6. $y = (3/4)x - 1$
7. $y = (-1/2)x + 8$
8. $y = (5/3)x + 2$
9. $y = 5x - 12$
10. $y = -3x + 4$

FINDING THE EQUATION OF A LINE USING POINT-SLOPE FORM

PAGE 38

1. $y = 3x - 3$
2. $y = 2x + 2$
3. $y = -3x - 3$
4. $y = -2x$
5. $y = 4x + 1$
6. $y = -4x - 14$

7. $y = 4$
8. $y = (1/2)x - 2$
9. $y = -x$
10. $y = (-3/2)x + 7$
11. $y = -4x - 4$
12. $y = (-2/3)x - 4$

FINDING THE EQUATION OF A LINE FROM A POINT AN THE Y-INTERCEPT

PAGE 39

1. $m = -1$ SI is $y = -x + 3$.
2. $m = 1$ SI is $y = x + 1$.
3. $m = -2/3$ SI is $y = (-2/3)x + 2$.
4. $m = 2$ SI is $y = 2x - 2$.
5. $m = -1$ SI is $y = -x + 1$.
6. $m = 3/4$ SI is $y = (3/4)x + 3$.

FINDING THE EQUATION OF A LINE FROM TWO POINTS

PAGE 40

1. $m = 2$ $b = 1$ SI is $y = 2x + 1$.
2. $m = 1$ $b = 3$ SI is $y = x + 3$.
3. $m = 2$ $b = -10$ SI is $y = 2x - 10$.
4. $m = -1/2$ $b = -2$ SI is $y = (-1/2)x - 2$.
5. $m = 3$ $b = 8$ SI is $y = 3x + 8$.
6. $m = -1$ $b = 3$ SI is $y = -x + 3$.
7. $m = 1/4$ $b = 3$ SI is $y = (1/4)x + 3$.
8. $m = -1/5$ $b = 24/5$ SI is $y = (-1/5)x + 24/5$.
9. $m = 3/2$ $b = 0$ SI is $y = (3/2)x$.
10. $m = 3/2$ $b = 0$ SI is $y = (3/2)x$.

PAGE 41

1. $y = -2x + 9$
2. $y = x + 4$
3. $y = -x - 4$
4. $y = x + 5$
5. $y = (1/2)x$
6. $y = 3x + 7$
7. $y = (1/2)x - 2$
8. $y = (2/3)x + 3$
9. $y = -x + 4$
10. $3x + y = 5$
11. $x - y = 2$
12. $x - 2y = -8$
13. $x + 3y = 0$
14. $x - 2y = 1$
15. $x + 3y = -3$
16. $2x - y = -10$

FINDING THE EQUATION OF A LINE WITH ZERO OR NO SLOPE

PAGE 42

1. $m = 0$ $y = 3$
2. no slope $x = -3$
3. no slope $x = 5$
4. no slope $x = -1$
5. $m = 0$ $y = -3$
6. $m = 0$ $y = -2$

MIXED REVIEW OF LINEAR EQUATIONS

PAGE 43

1. $y = 3x - 5$
2. $y = -2x + 3$
3. $y = -x + 6$
4. $y = (1/2)x - 1$
5. $y = (-4/3)x + 5/2$
6. $y = 3x - 4$
7. $y = -2x + 5$
8. $y = 5x + 10$
9. $y = (3/4)x - 7$
10. $y = (-1/2)x + 6$
11. $y = -x + 3$
12. $y = x - 3$
13. $y = -2x - 1$
14. $y = (1/2)x - 4$
15. $y = (-3/2)x$
16. $y = 5$
17. $x = -1$ (vertical line)
18. $y = (2/3)x - 2$

FINDING THE EQUATION OF A LINE FROM A GRAPH

PAGE 44

$m = 1/2$ $b = 1$ Equation: $y = (1/2)x + 1$

PAGE 45

1. $y = 2x - 2$
2. $y = (1/2)x - 1$
3. $y = -3x + 2$
4. $y = (2/3)x - 2$
5. $y = -x - 1$
6. $y = (-1/2)x + 1$

GRAPHING LINEAR EQUATIONS

PAGE 46

1.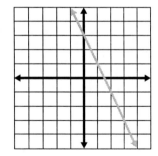

2.

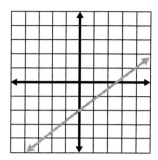

6.

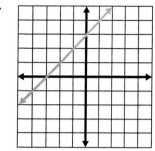

3.

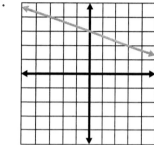

7.

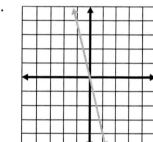

4.

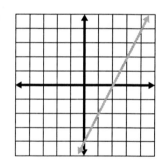

8.

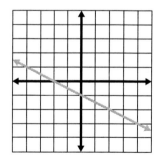

5.

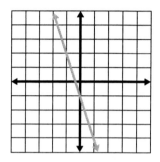

PAGE 47

1.

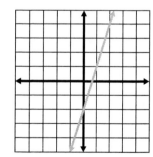

2.

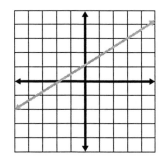

6.

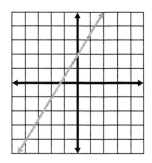

3.

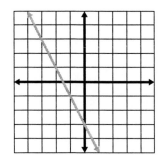

7.

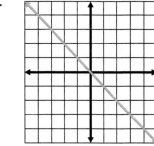

4.

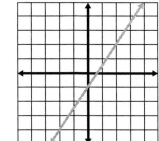

8.

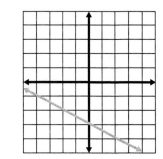

5.

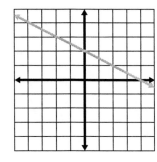

*G*RAPHING *S*YSTEMS OF *E*QUATIONS
*P*AGE 48

1. (2,4)   2. (3,–2)

*G*RAPHING *P*ARALLEL AND *P*ERPENDICULAR *L*INES
*P*AGE 49

1. (2,0)   2. Parallel—no solution

MIXED REVIEW OF GRAPHING LINEAR EQUATIONS

PAGE 50

1.

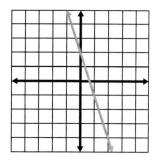

2.

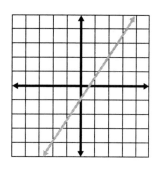

3.

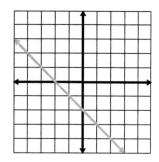

4.

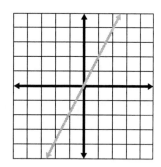

5.   Solution: (−1,4)

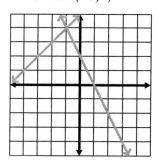

6.   Parallel—no solution

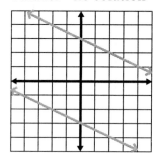

GRAPHING LINEAR INEQUALITIES

PAGE 51

1.

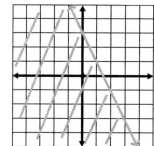

2.

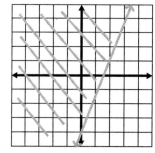

PAGE 52

1.
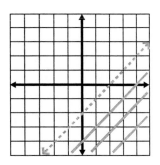

2.
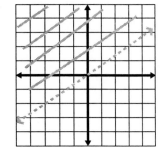

3.

4.
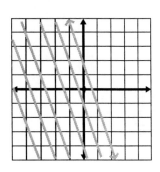

PAGE 53

1.
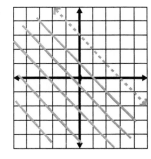

2.
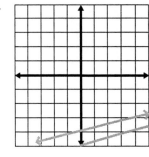

3.
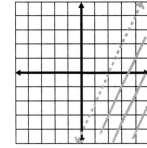

4.
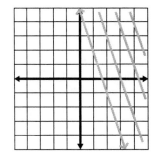

5.

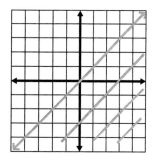

6.

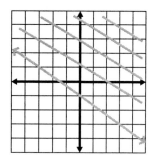

7.

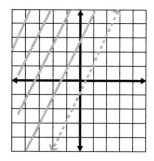

8.

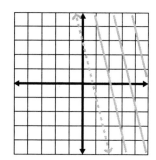

GRAPHING SYSTEMS OF LINEAR INEQUALITIES

PAGE 54

1.

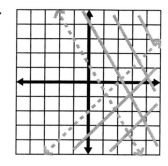

2.

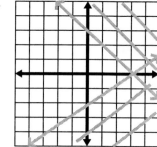

3.

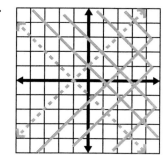

4.

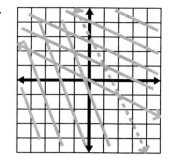

GRAPHING DATA POINTS AND FINDING THE SLOPE

PAGE 55

Slope: 2.2

Meaning: Conversion rate from kg to pounds

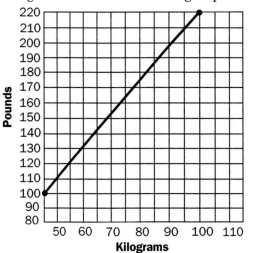

PAGE 56

1.

Time (sec)	Distance (ft)
1	12
2	24
3	36
4	48
5	60
6	72
7	84

Velocity = 12 ft/sec

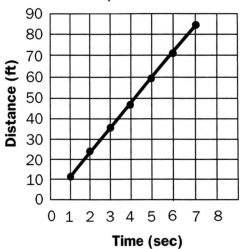

2.

Force (newtons)	Acceleration (m/s²)
20	5
40	10
60	15
80	20
100	25
120	30
140	35
160	40

Mass = 4 kg = 8.8 lbs

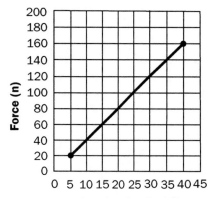

PAGE 57

3.

Time (sec)	Distance (ft)
0	0
5	400
10	800
15	1200
20	1600
25	2000
30	2400

Velocity = 80 ft/sec

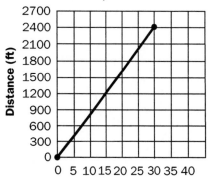

4.

Force (newtons)	Acceleration (m/s^2)
0	0
45	1
90	2
135	3
225	5
315	7
360	8
405	9

Mass = 45 kg = 99 lbs

This mass could be a small person.

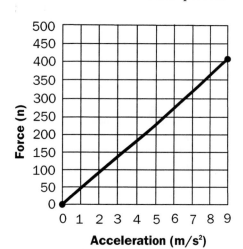

GRAPHING QUADRATIC EQUATIONS

PAGE 58

	x-intercept	y-intercept	vertex
1.	(−1,0) (4,0)	(0,−4)	(1.5, −6.25)
2.	(2,0) (−2,0)	(0,−4)	(0,−4)
3.	(1,0) (−1,0)	(0,1)	(0,1)

PAGE 59

	x-intercept(s)	y-intercept	vertex
1.	(0,0)	(0,0)	(0,0)
2.	(−3,0) (1,0)	(0,−3)	(−1,−4)
3.	none	(0,−9)	(0,−9)
4.	(1,0)	(0,1)	(1,0)
5.	(2,0) (−2,0)	(0,4)	(4,0)

1.

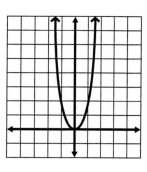

2.

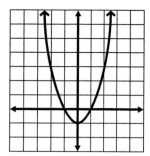

3.

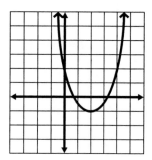

4.

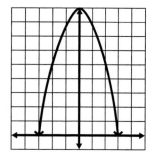

PAGE 60

1. *x*-intercept(s):   none
    y-intercept: (0,1)
    vertex: (0,1)

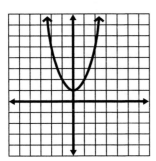

3. *x*-intercept(s):   (1,0)
    y-intercept: (0,1)
    vertex: (1,0)

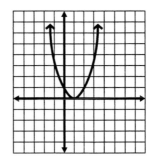

2. *x*-intercept(s):   (–1,0) (1,0)
    y-intercept: (0,2)
    vertex: (0,2)

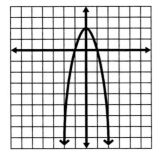

4. *x*-intercept(s):   (–4,0) (1,0)
    y-intercept: (0,–4)
    vertex: (–1.5,–6.25)

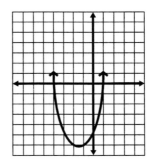

GRAPHING CIRCLES

PAGE 61

1. Radius: 3   Center: (1,2)

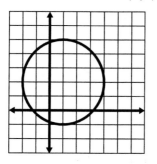

2. Radius: 5   Center: (0,0)

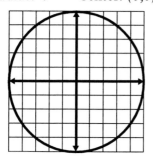

3. Radius: 2   Center: (−2,−1)

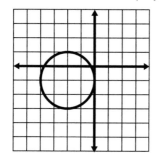

4. Radius: 4   Center: (1,−3)

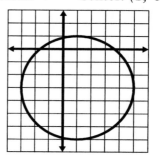

PAGE 62

1. Radius: 4   Center: (1,2)

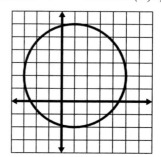

2. Radius: 3   Center: (0,4)

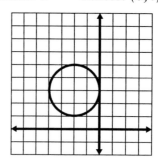

3. Radius: 2   Center: (–2,3)

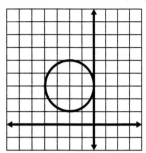

4. Radius: 1   Center: (–1,–1)

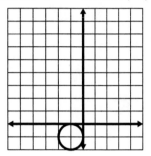

5. Radius: 2   Center: (4,–2)

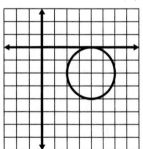

6. Radius: 3   Center: (–4,0)

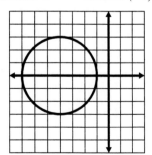

PAGE 63

1. Radius: 3   Center: (3,1)

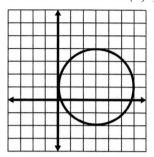

2. Radius: 4   Center: (0,0)

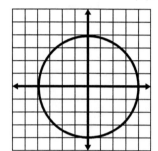

3. Radius: 2   Center: (2,–2)

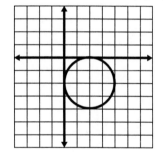

4. Radius: 4   Center: (–3,–2)

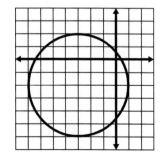

5. Radius: 5   Center: (1,1)

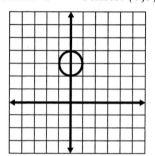

2. Points: (5,0) (–5,0) (0,2) (0,–2)

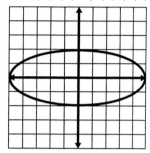

6. Radius: 1   Center: (0,3)

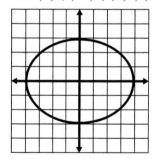

3. Points: (2,0) (–2,0) (0,3) (0,–3)

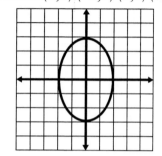

GRAPHING ELLIPSES

PAGE 64

1. Points: (4,0) (–4,0) (0,3) (0,–3)

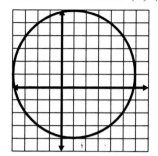

4. Points: (5,0) (–5,0) (0,4) (0,–4))

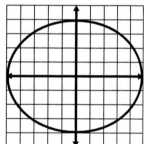

PAGE 65

1. Center: (3,1)
   Horizontal vertices: (6,1) (0,1)
   Vertical vertices: (3,6) (3,–4)

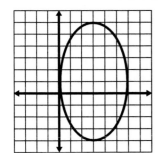

2. Center: (–2,2)
   Horizontal vertices: (0,2) (–4,2)
   Vertical vertices: (–2,6) (–2,–2)

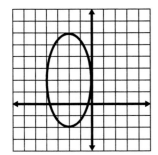

PAGE 66

1. Center: (0,0)
   Horizontal vertices:(3,0) (–3,0)
   Vertical vertices: (0,5) (0,–5)
   Major axis: 10
   Minor axis: 6
   Foci: (0,4) (0,–4)

2. Center: (–3,1)
   Horizontal vertices: (7,1) (–13,1)
   Vertical vertices: (–3,9) (–3,–7)
   Major axis: 20
   Minor axis: 16
   Foci: (3,1) (–9,1)

GRAPHING HYPERBOLAS

PAGE 67

1. Center: (2,3)
   Foci: (7,3) (–3,3)
   Opens left and right.

2. Center: (0,0)
   Opens down left and up right.

3. Center: (–3,1)
   Foci: $(-3,1 + \sqrt{5})$ $(-3,1 - \sqrt{5})$
   Opens up and down.

PAGE 68

1. Center: (0,0)
   Vertices: (2,0) (–2,0)
   Foci: $(2 + \sqrt{29},0)$ $(2 - \sqrt{29},0)$

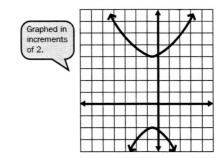

2. Center: (–1,2)
   Vertices: (–1,8) (–1,–4)
   Foci: (–1,12) (–1,–8)

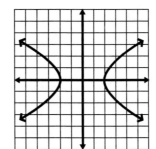

Graphed in increments of 2.

3. Center: (0,0)

   Vertices: (−5,5) (5,−5)

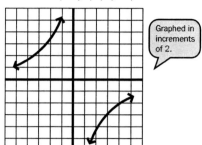

Graphed in increments of 2.

4. Center: (−1,−3)

   Vertices: $-1 - \sqrt{2}, -3$ $-1 + \sqrt{2}, -3$

   Foci: (−3,−3) (1,−3)

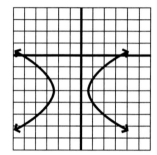

PAGE 69

1. Center: (0,0)

   Vertices: (−2,0) (2,0)

   Foci: $(-\sqrt{41}, 0)$ $(\sqrt{41}, 0)$

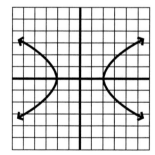

2. Center: (−1,2)

   Vertices: (−1,5) (−1,−1)

   Foci: (−1,7) (−1,−3)

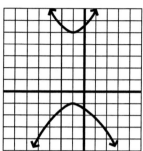

3. Center: (0,0)

   Vertices: $(2\sqrt{5}, 2\sqrt{5})$ $(-2\sqrt{5}, -2\sqrt{5})$

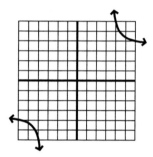

4. Center: (4,−1)

   Vertices: $(4 + \sqrt{3}, -1)$ $(4 - \sqrt{3}, -1)$

   Foci: (7,−1) (1,−1)

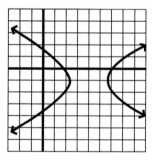

PRACTICAL APPLICATIONS

PAGE 70

1. 400 miles ÷ 8 hours = 50 mph average

2. $x^2 + y^2 = 9$, $x^2 + y^2 = 36$, $x^2 + y^2 = 81$, $x^2 + y^2 = 144$, $x^2 + y^2 = 225$

3. $s = (1/2)gt^2$ so $s = (1/2)(9.8 \text{ m/s}^2)(10^2) = $
490m

4. $s = (1/2)gt^2$ so $315 = 1/2(9.8)t^2$ so $t^2 = $
64.286 so $t = 8.0$ sec

5. Graph the data points

Time (sec)	Height (ft)
0	0
1	80
2	128
3	144
4	128
5	80
6	0

Max height = 144 ft

Total time = 6 sec

PAGE 71

1. $1000 = 1/2(9.8)t^2$ ⟹ $1000 = (4.9)t^2$ ⟹
$204.08 = t^2$ ⟹ $t \approx 14.3$ sec

2. Graph the data points

Time(s)	Height (ft)
0	0
1	224
2	416
3	576
4	704
5	800
6	864
7	896
8	896
9	864
10	800
11	704
12	576
13	416
14	224
15	0

Max height = 900 ft   Total time = 15 sec

3. $16t^2 - 32t - 240 = 0$
$16(t^2 - 2t - 15) = 0$
$(t - 5)(t + 3) = 0$
$t = 5, t = -3$
t must be positive, so $t = 5$ sec

4. $0.25x^2 - 25x + 600 = 0$
$0.25(x^2 - 100x + 2400) = 0$
$x^2 - 100x + 2400 = 0$
50 wheels

5. $-0.0025x^2 - 100x = 0$
$-0.0025(x^2 - 40,000x) = 0$
$x^2 - 40,000x = 0$
$x(x - 40,000) = 0$
$x = 0$, or 40,000
max is 20,000

MIXED REVIEW OF GRAPHS

PAGE 72

1.

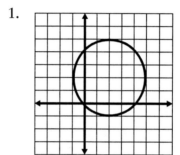

2.

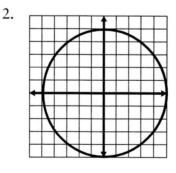

3.

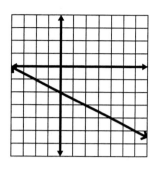

1.

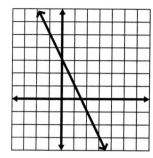

4.

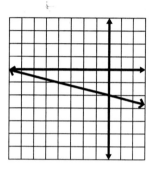

2.

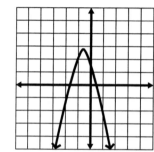

5.

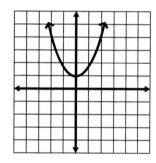

3.

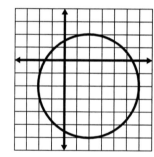

6.

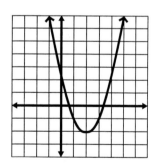

4.

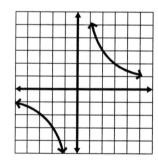

5.

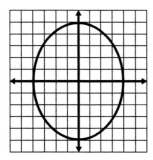

6.

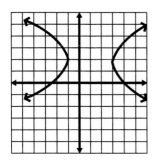

BAR GRAPHS

PAGE 75
1.

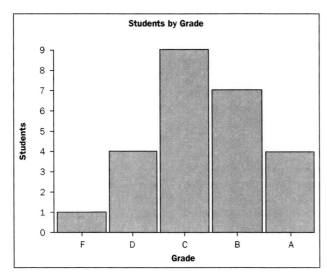

2.

Rainfall is heaviest in the winter.
Rainfall is lightest in the summer.

PAGE 77

3.

The substances closest in value are blood, pure water, and sea water. These substances are mostly water.

HISTOGRAMS

PAGE 79
1.

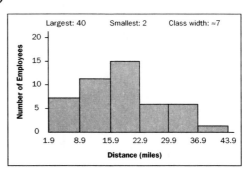

PAGE 80

2.

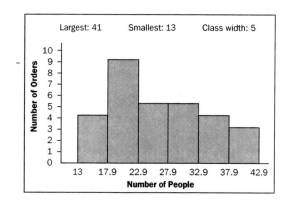

PAGE 81

3.

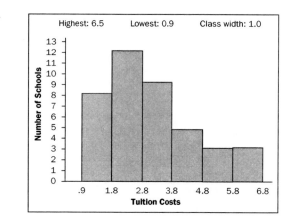

PAGE 82

4.

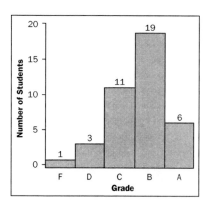

LINE GRAPHS

PAGE 84

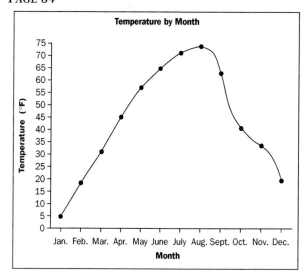

The months to be avoided are December, January, February, March and maybe November.

PAGE 85

1.

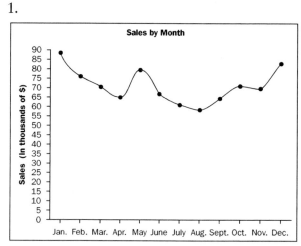

PAGE 85

PAGE 85

2.

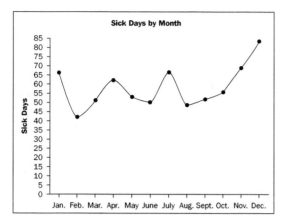

The four months with the most sick days are November, December, January, and July. July does not seem to fit with the other months. It is not known for colds and flu, like the winter months, but it may be a time of year when allergies are particularly bad. Another possibility is that workers call in sick in July in order to enjoy the warm weather.

PIE GRAPHS

PAGE 87

1.

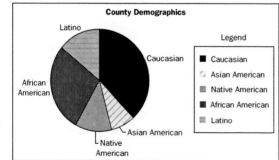

PAGE 88

2.

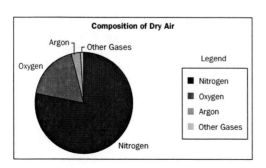

3.

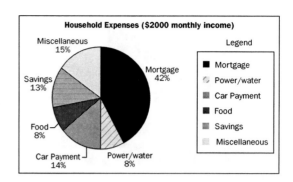

Carl spends 14 percent of his income on his car each month.

4.

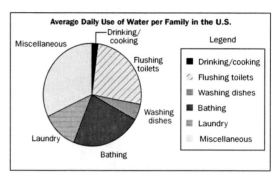

DUAL LINE GRAPHS

PAGE 89

1.

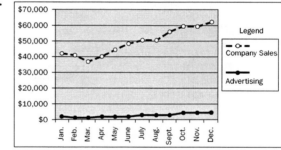

2. Sales go up as advertising goes up.
3. No. Pie graphs do not show growth in the same way as a line graph.
4. Sales would continue to increase.

Share Your Bright Ideas with Us!

We want to hear from you! Your valuable comments and suggestions will help us meet your current and future classroom needs.

Your name_____Date_____

School name_____Phone_____

School address_____

Grade level taught_____Subject area(s) taught_____Average class size_____

Where did you purchase this publication?_____

Was your salesperson knowledgeable about this product?   Yes_____   No_____

What monies were used to purchase this product?

___School supplemental budget   ___Federal/state funding   ___Personal

Please "grade" this Walch publication according to the following criteria:

Quality of service you received when purchasing .... A   B   C   D   F
Ease of use.. A   B   C   D   F
Quality of content.. A   B   C   D   F
Page layout .... A   B   C   D   F
Organization of material .... A   B   C   D   F
Suitability for grade level .... A   B   C   D   F
Instructional value... A   B   C   D   F

COMMENTS:_____

_____

What specific supplemental materials would help you meet your current—or future—instructional needs?

_____

Have you used other Walch publications? If so, which ones?_____

May we use your comments in upcoming communications?   ___Yes   ___No

Please **FAX** this completed form to **207-772-3105**, or mail it to:

Product Development, J.Weston Walch, Publisher, P.O. Box 658, Portland, ME 04104-0658

We will send you a **FREE GIFT** as our way of thanking you for your feedback. **THANK YOU!**